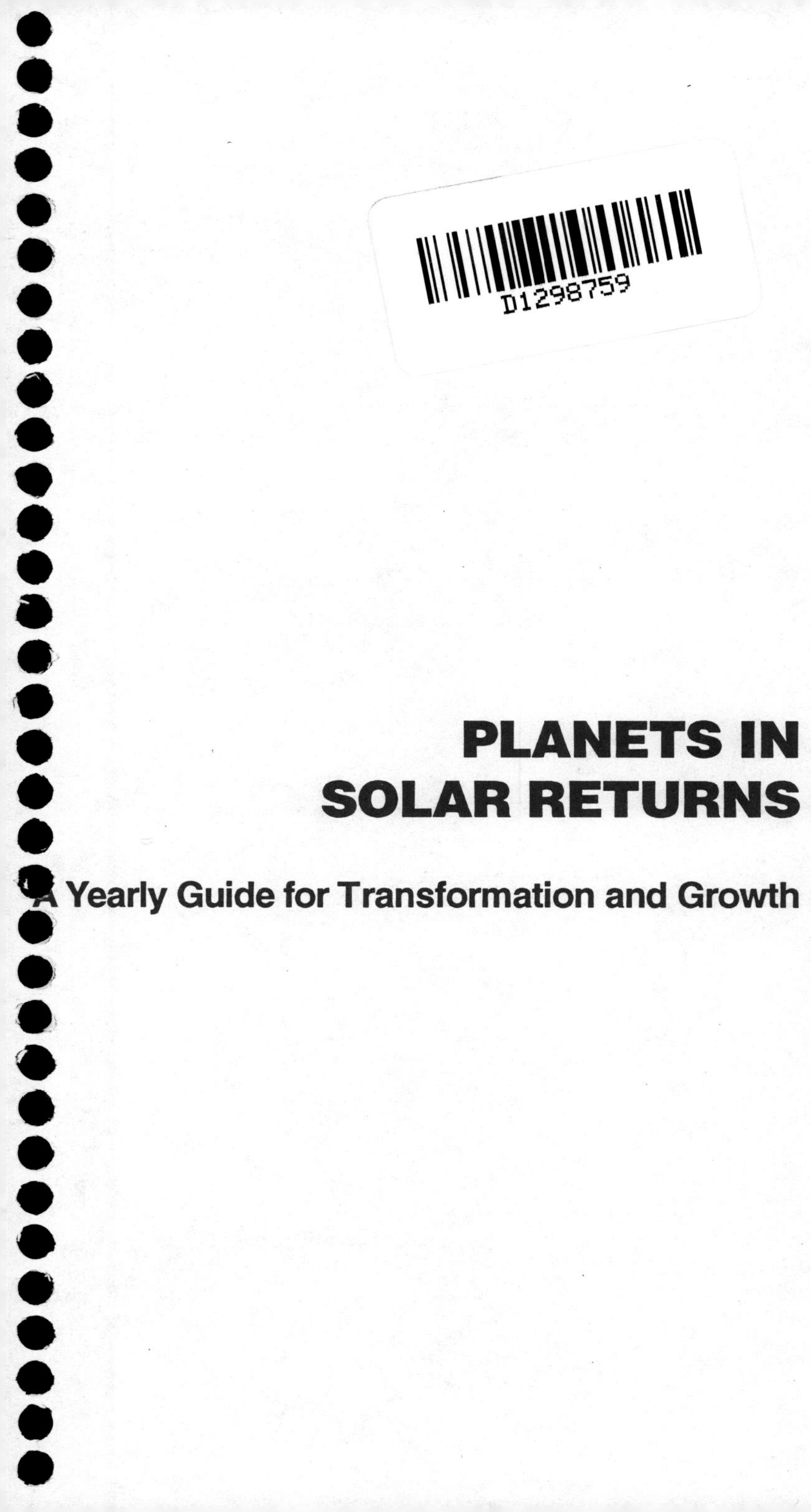

PLANETS IN SOLAR RETURNS

A Yearly Guide for Transformation and Growth

Planets in Solar Returns

A Yearly Guide for Transformation and Growth

Mary Shea

Cover designed by Maria Kay Simms

ISBN 0-917086-96-1

Printed in the United States of America

Published by ACS Publications, Inc.
5521 Ruffin Road
San Diego, CA 92123

First Printing, March 1992
Second Printing, October 1992
Third Printing, January 1995

With Love to Mom and Dad
who guided my earliest years of growth.

ACKNOWLEDGMENTS

Many people played a crucial role in the development and writing of this book. Without their assistance this project never would have been completed. I am deeply grateful to the following individuals for their help, support, understanding, expertise, and love:

my husband, Bob, for years of assistance on the home front. Bob not only proofread the original chapters, but he also cooked, cleaned and cared for children while I worked.

Joan and Ken Negus, for their personal encouragement and professional dedication to astrology.

students Karen, Lillian, Micheline, Marge, Nancy, Paula, Ruth, and Shirley for their questions, answers, insights, laughter, and quest for knowledge. I remember the good times.

all my clients and friends for giving me valuable feedback on the material, especially Paula and Beverly.

Neil Michelsen and Maritha Pottenger for teaching me to integrate my philosophy and interpretations into a consistency of thought, word and action.

last, but by no means least, Mary Jo Putney, an excellent writer and good friend for proofreading my book, polishing my style, correcting my errors and talking me through the crazy days of rewrites.

Contents

CHAPTER FIVE

CHAPTER SIX

CHAPTER SEVEN

CHAPTER EIGHT

CHAPTER ONE

INTRODUCTION TO SOLAR RETURNS AND THEIR INTERPRETATION

What is a Solar Return Chart?

A solar return chart is a chart erected for the time that the transiting Sun returns to the position of the natal Sun. Approximately once every year the transiting Sun goes through the entire zodiac, every degree, minute and second of each sign. When you are born, the Sun in your natal chart has a specific position in the zodiac. This position can be measured exactly. At some point in time during each subsequent year, the transiting Sun returns to this natal position, conjuncting your natal Sun. The date and time when the conjunction occurs down to the exact second of arc is the time of the solar return chart. A chart is then calculated for that date, using the time (of the transiting Sun's return to your natal Sun's position) and your location (longitude and latitude) at the time of the conjunction.

Tropical, Placidus, Non-precessed Solar Return

There are many different kinds of solar return charts: tropical, sidereal, precessed and non-precessed. This book is based on research with the tropical, Placidus, non-precessed solar return chart. Interpretations are very similar to natal interpretations and this should make the material more understandable to both the professional and amateur astrologer. Basically, if you can read a natal chart, you can understand a solar return. No charts are needed other than the one solar return for the year.

Solar Return Location

The solar return chart is generally calculated for your location at the time of the Sun's return. This may occur on your birthday, or the day before or after. During leap years, it sometimes occurs two days before your birthday. If you are going to be away or traveling near your birthday, it is a good idea to calculate the time of the Sun's return before you leave; then you can note your position at the exact moment of the solar return.

You can also calculate the chart for your natal birth location even though you no longer live there. The natal location solar return chart is not as definitive as the relocation solar return, but it can give you good information. Some individuals prefer the chart calculated for their place of residence regardless of where they are located at the time of the Sun's return. This chart can be useful also, but again, it tends to be secondary to the chart of your actual location at the time of the Sun's return to its natal zodiacal position.

Some individuals relocate for a short period of time to adjust the house placements of the planets in the solar return. Those who calculate the chart for where they are at the time of the transiting Sun's conjunction to the natal position sometimes travel to other locations to change the yearly interpretation. The relocation adjusts the angles and cusps, orienting the planets into different houses. The zodiacal degrees and aspects of the planets remain the same, but because of the different planetary placements different areas of life are emphasized.

Solar Return Calculations

There are several different ways to calculate a solar return chart and it is always imperative that the seconds on the natal Sun's position be taken into consideration. The Sun moves approximately $2\,^1/_2$ minutes per hour or a half of a minute of arc every 12 minutes of time. If you do not include the natal Sun's exact seconds in your calculation, the rounded off Sun position could distort the time of the Sun's return. Small changes in the Sun's position can translate into bigger changes in the solar return chart. The Midheaven could be off 3-4 degrees and your Ascendant could be off as much as 8 degrees in signs of short ascension. The technique you use and the accuracy of your calculations are very important.

Two widely used calculation methods tend to produce inaccurate charts. Do not use logarithms to calculate your solar returns; log tables are inaccurate when one is working high in the first column where the differences are great. Solar returns are always calculated in the first column. Also, do not use age regression tables appearing in some books. These tables are based on the average chart progression. The average chart only occurs twice a year and your chances of being average are 2 in 365. Computer computation is probably the best route for those who are not mathematically inclined.

How to Tell If Your Calculations Are Correct

The solar return chart Midheaven has a thirty-three year cycle. This means that if you are living at your birth location when you are thirty-three years old, your solar return angles will be roughly the same as your natal angles. There is a cycle of rotation evident in solar returns. Each year the Midheaven of the solar return chart moves forward three signs, plus or minus a few degrees. This fact will help you to determine if you have calculated your solar returns correctly.

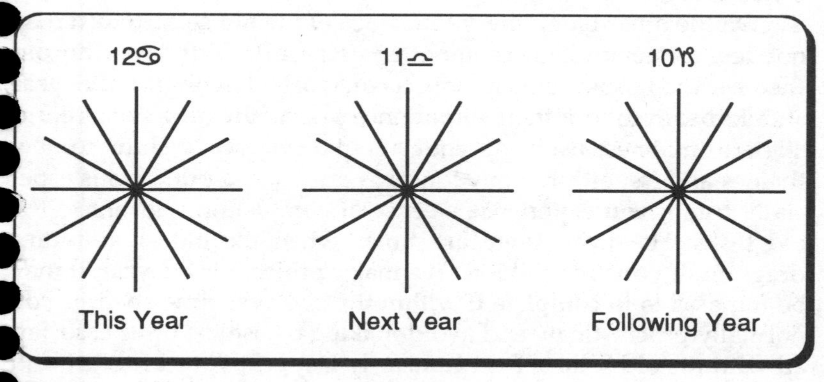

Figure 1

For example, if your Midheaven is 12 degrees Cancer this year, next year your Midheaven could be approximately 11 degrees Libra and the year following it might be 10 Capricorn (assuming that you remain in the same location).The Midheaven will continue to move forward three signs while also shifting plus or minus a few degrees, year after year, until it has eventually moved forward and shifted back to your natal Midheaven. (See figure 1.)

Period of the Solar Return's Significance

The significance of the solar return runs birthday to birthday with a three month overlap at the beginning and end of each year. The symbolism of the new solar return can be felt as much as three months before the birthday by very intuitive people. Usually, at this time, one becomes aware of new directions, opportunities and problems. Many times there is even an event exactly three months before the birthday which triggers awareness and signals the beginning of the new solar return. This event is usually very much related to the new solar return and very much out of context with the old one. If you have plans to travel overseas during the three months before your birthday, do not look for the event in the old solar return even if you have been planning to take the trip all year. The old solar return will probably indicate the planning stage, but the new solar return will indicate the trip. The trip itself illustrates change, and since it occurs in the three months before your birthday, it will most likely be indicative of the new solar return chart.

On the other hand, the significance of the old solar return may not feel passé until three months after your birthday. The old themes and issues which you have worked with for the year should begin to lose their importance just as the new solar return starts to manifest itself. As your attention naturally shifts to new themes and issues, previous concerns are phased out. This is especially true if you experience a sense of completion relevant to the old tasks. However, there are times when the old solar return drags on. If you are working on a major project for the year, it may be impossible to complete it within the one year time span. If you normally procrastinate and avoid making decisions, tasks associated with the old solar return can easily last past your birthday and into the first three months of the new solar return year. Unresolved issues carried over from year to year become stumbling blocks to further advancement and development. Individuals who consistently avoid facing the real issues in their lives accumulate major problems which are very difficult to handle.

The main transitional month for the change from the old solar return to the new solar return is the month directly preceding the birthday. Issues related to the old solar return are resolved at this time unless they drag on (as explained above), or they are part of the new solar return as well. Issues related to the new solar return become more pressing during this transitional month and are certainly full blown by the birthdate. Sometimes the new solar return simply gains strength during the three months before your birthday and sometimes it comes in with a bang on or near your birthday. Rarely does it happen that a solar return chart starts to change manifestation later in the year. Themes in the chart usually persist for the entire year, but once in a great while one new theme will appear later than expected. In every case, the new theme can be seen in the solar return chart, yet it lay there inactive for some unknown reason. An external event is the triggering mechanism used to activate this inert awareness, and the event generally occurs within the three months following the birthday. It can be hypothesized that unconscious factors help to avoid full involvement with this new issue until such time as it is associated with an external event of importance and thereafter becomes very obvious.

Generally, the greater the difference between the interpretation for the new solar return chart and the interpretation for the old solar return chart or the natal chart, the greater the difficulty in adjusting to the new issues and the more unsettled the transitional period in the month before your birthday.

☉☿♀♂♈ ☍ ♃♄♇♎

For example, in the Spring of 1981, Jupiter, Saturn and Pluto were all in the sign of Libra while all the personal planets, Mercury, Venus, and Mars were travelling with the Sun in Aries. The solar returns for Aries individuals all had a strong Libran emphasis with a number of oppositions forming an hourglass chart pattern. Aries individuals are more focused on self-development and self-interests than they are on others and relationship needs; however, in 1981, most of them experienced a deep concern for the "significant other" in their lives. This shift in awareness caught many Aries individuals by surprise and they felt that they were not adequately prepared for this emphasis on others or for the decisions they needed to make concerning their relationships. Their lack of experience with these issues delayed their awareness.

One woman's father developed Alzheimer's disease and she was forced to commit him after he tried to attack her. She was unaware of his violent tendencies until they were a physical reality. Another woman left her home of many years to move across the country with her husband only to be told that he wanted a divorce. She was unaware of his ambivalent feelings towards the marriage. A male client was seriously beaten by his live-in lover. He was unaware of his lover's anger. Strong theme changes which are indicated by strong changes in the solar return chart can reveal inexperience and ineptitude in those individuals caught unaware and unprepared.

Some Aries individuals experienced 1981 as a very positive and productive year. Being aware of their increased need for fulfillment in relationships, they focused on issues that blocked greater intimacy or threatened communication. They had the courage to tactfully discuss sensitive topics such as fidelity, dependency, and death. They willingly worked with situations and issues involving significant others rather than trying to escape. By being more aware of the needs of others, they were able to combine their efforts with those they loved to form a united front and a supportive structure. Their relationships grew and deepened as the issues they faced became easier to handle.

Judging differences between a new solar return and an old one, or a solar return and a natal chart, applies to elements, modes and hemispheric emphasis as well as general interpretation. It is difficult for an air person to adjust to a watery Cancerian solar return. If there is discomfort, it will most likely be experienced during the transitional month before the birthday, but sometimes the individual feels uneasy during other parts of the year. As the more diverse issues of the year are integrated and worked with, one begins to function comfortably with the themes of the new solar return.

Introduction to Interpretation

The solar return chart is very simple to interpret and it is probably the easiest predictive technique to use. The most important thing to remember about the interpretation of the solar return chart is that it can be treated like a natal. If you can interpret a natal chart, then you can interpret a solar return. None of your expertise in natal interpretation will be lost. Almost anything you do to a natal can also be done to a solar return. Astrologers have successfully used hemispheric emphasis, elements, modes, asteroids, Sabian symbols, and general interpretation to understand solar returns. Probably almost any astrological technique will yield an accurate interpretation as long as you are willing to work with the technique long enough to become familiar with the manifestation. Observation is the key to understanding new applications.

The major difference between solar return interpretation and natal interpretation is the matter of time. The solar return is a temporary chart, and therefore the action unfolds and proceeds more quickly. You are not dealing with a life-long pattern, but a manifestation that lasts for approximately one year. Interpretations are more concentrated since time is compressed. What is seen as a drawn out or intermittent tendency in the natal chart, becomes a prominent issue or theme in the short span of the solar return.

Reading the solar return chart like a temporary natal may be a difficult task when you are first starting out. For this reason, much of this book focuses on interpretation. Narrow interpretations of singular astrological data are never the ideal. It is difficult to isolate one factor for interpretation without considering the rest of the chart. The interpretations presented here should be used as guidelines only, and should be modified according to contraindications in the chart.

The Solar Return Script

The solar return chart is the birth chart of a new year in your life. Each year you are faced with new issues as different areas of your life become dominant. The solar return is like a script for a year-long play and you get to play the lead. The script has a beginning, a middle and an ending. It also has a theme, a plot, several subplots, a twist or two, and a moral to the story (also known as the spiritual interpretation). The theme is well defined and usually it is easy to identify with. Each year tasks need to be accomplished, opportunities need to be taken advantage of, conflicts need to be handled, talents need to be developed, and growth should occur. All of these changes are written into the script, but allowances are made for a great deal of improvisation. The planetary positions show the areas of interest and prominent activity. The aspects show areas of support or conflict of interests. The themes for the year are reflected as a cohesive whole in the interpretation of the entire solar return chart. They are seen as the motivating psychological force behind the experiences of the year.

The solar return is a better predictor of psychological change and internal growth than it is of actual physical events. This is because the chart seems to be more closely related to psychological influences and patterns of growth than it is to the events which allow these influences to manifest themselves. Events appear to be of secondary importance and can be mere vehicles for the expression of feelings already present and needing to be experienced. Events are not crucial to the growth process and are therefore harder to predict. As with any chart, it is difficult, if not impossible, to assess the individual's level of evolution and ability to express free will. Nobody can predict how well one will handle the feelings shown in the solar return or determine whether the year will be an easy one or a difficult one. Choosing to work positively with the themes of your solar return can make life seem more understandable and directed. Ignoring or avoiding the lessons of your solar return can result in increased difficulties and a subsequent crisis situation which could have been avoided had you faced certain issues earlier. Consequently, it is far easier to predict the feelings and thoughts associated with the solar return than it is to predict the exact events to transpire.

Perhaps an example will help to clarify the importance of the psychological influences behind actual events. When Saturn is in the 4th house of the solar return chart, you will most likely feel more responsibility towards your home or family. The events which will allow that feeling to manifest itself are very diverse. You may have a grown son or daughter move back into your house with small children from a broken marriage. Your aging mother or father may depend on you for help and guidance. You might decide to move to a very old home and restore the original decor. The underlying theme is increased responsibility within the home and family, and exactly which event or events allow you to express that feeling is secondary.

Although solar return charts primarily reflect psychological feelings and motivations, their manifestations can occur on many levels. Interpretations are not limited to the psychological level, but are reflected in events, involvement in physical situations, health, emotional feelings, and spiritual considerations. Using the example of Saturn in the 4th house again, we can see that the emphasis on responsibility can be realized in several ways. The person with Saturn in the 4th house would naturally concentrate energy on an event that entailed extra responsibility and more work in the home. In the process, it is possible to feel overburdened, lonely or isolated. Physically, the extra responsibility and increased activity might be draining, leaving little energy or enthusiasm for pleasure. Spiritually, the individual might be guilt-laden. But it is just as likely that the native will develop a new sense of stability and security that grows from a dedication to the needs of family members and domestic situations. Rather than feeling drained, this person could become more organized and accomplished. He or she may feel a renewed sense of spiritual purpose while laying the groundwork for endeavors to come.

The House Placement of a Planet

The house placement of a solar return planet is more important than the sign it is in. There are obvious reasons for this. The Sun is always in the same sign and degree; that is the nature of the solar return chart. Because of the eclipse cycle, the Moon is limited to nineteen placements in the solar return chart. After nineteen years, the Moon begins to repeat itself; therefore, the lunar placements are limited and repetitive. Mercury can only be at most one sign away from the Sun, and consequently, the placement of Mercury is limited. Venus has only eight placements in the solar return. If you do eight solar returns in a row, you will have all eight of your solar return Venus placements for your lifetime. On the ninth year, the Venus placement will begin to repeat itself, usually within a degree. In fact, if you look at these placements and how they aspect your natal planets, you will discover a lot of information about your love life and why your relationships feel more comfortable some years than others.

Saturn, Uranus, Neptune and Pluto are slow moving planets, and are usually in the same sign in everyone's solar return chart during any given year. That leaves Jupiter and Mars. Jupiter spends approximately one year in a sign and chances are you will experience a new sign every year. Reason tells us that Jupiter's sign cannot possibly be important when everybody else has it; however, experience has shown that Jupiter's sign can be important when Jupiter is in aspect to the Sun, Moon or in a major configuration such as a T-square, Grand Cross or Grand Trine. In these situations, it is worth noting Jupiter's sign. Mars, and possibly the Moon, are the only other planets that can have significance by sign. As a general rule, concentrate your interpretation on the house placement of the planet rather than the sign.

Planets conjunct a house cusp within a few degrees may be read in both houses. It is especially important to read outer planets in both houses since they eventually transit or retrograde into the other house, indicating dual or shifting concerns during the year.

The more planets that are in a solar return house, the more emphasized that area of life will be during the coming year, especially if the Sun is one of the planets present. For example, the Sun in the 7th house of the solar return indicates that relationships and the needs of others are very important during the coming year. When other planets also fall into the 7th house, the emphasis tends to increase with each addition. Four or more planets in any one house indicate a strong need to be involved with those themes and issues. But the individual may become so focused on this one area of life that perception is distorted and these themes are overemphasized. He or she may not want to think about anything else or accomplish tasks in other areas. The overloaded house becomes a symbol for obsessive preoccupation.

An individual with four planets in the 7th house gave up his home, job, and way of life to move closer to a romantic partner. Only the lover mattered and he was willing to make major changes to maintain the relationship and accommodate the needs of his loved one. His fascination turned to disillusionment when the Sun and several other planets moved to the 4th house the following year. The individual realized that the cherished relationship was not meeting his needs and he began to express his lack of emotional fulfillment. One year later, the Sun and several planets rotated into the 1st house, and the man broke off the relationship and set

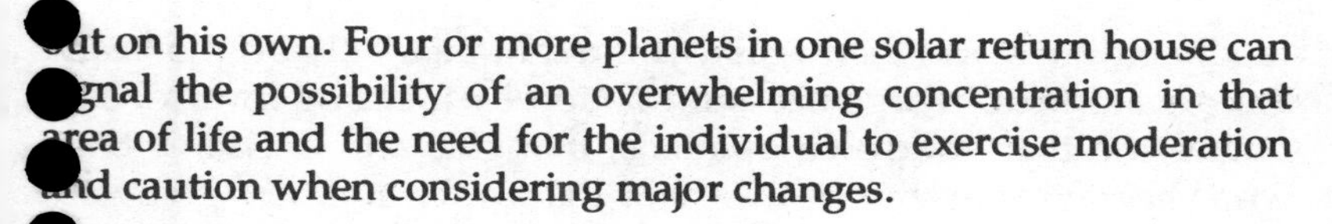

out on his own. Four or more planets in one solar return house can signal the possibility of an overwhelming concentration in that area of life and the need for the individual to exercise moderation and caution when considering major changes.

Retrograde Planets

Do not pay attention to retrograde outer planets in the solar return. There is about a 50/50 chance that the outer planets will be retrograde since they are each retrograde for almost six months out of the year. Generally, if an outer planet is positioned ninety degrees or more after your Sun's zodiacal placement, and not less than ninety degrees before your Sun's placement, it will be retrograde. Half of the population will have a retrograde Pluto in their solar returns this year. Some individuals will have Pluto retrograde in their solar return charts throughout their entire lifetime. Therefore, consider the retrogradation of Jupiter, Saturn, Uranus, Neptune and Pluto meaningless in the broad context of solar return interpretation. The interpretation of retrograde Mercury, Venus and Mars, however, is very important.

MERCURY RETROGRADE

Mercury is usually retrograde in the solar return chart every six years. When Mercury is retrograde, it is time to be introspective, especially about those things related to Mercury's house placement. You should be doing a lot of your own thinking and learning rather than depending on others. Integrate previously acquired information into your own individualized mind-set. You probably already know everything you need to know to handle a certain life situation. If you continue to depend on others for advice, you will find conversations meaningless within the context of your own intellectual needs. Your mental processes are not very receptive to new information. Your mind is like a cup that is filled to the brim. Any additional knowledge spills over and is lost.

You can experience this retrogradation as a certainty that you know you are right and only you can make the best decisions concerning your own future. What others tell you might clash with what you already know and you could tend to disregard their comments. You may be right, but the danger is that you may be wrong. You can be so in tune with your own thoughts that you are totally on the mark; then again, you might be totally off the beam. Take the time to reorganize information by focusing inward. You might find that your own opinions, thoughts and decisions truly work best for you. But be aware of the feedback others give you, which may be particularly valuable if you have missed the mark.

Secretiveness is also associated with Mercury retrograde. There is a tendency to withhold information and sometimes lie. Generally, there are two major reasons for doing this: one is that you really do have secrets which need to be kept; the other is that expressing your opinions or thoughts openly causes tension in your relationships. Retrograde Mercury is associated with biting your tongue and swallowing your own words in order to keep peace. Your true opinions may not emerge until the start of the next solar return.

This time is excellent for putting your thoughts down on paper and writing original material. You will be able to see things differently when your thoughts are written down and this is a good way to get organized. Old opinions and beliefs may be outdated so you need to reorganize your thinking and shed new light on a subject area.

VENUS RETROGRADE

Of the eight placements Venus has in the solar return chart, one is usually consistently retrograde, though irregularities occur. Venus retrograde indicates a time of comparison and contrast. Your increased ability to distinguish and value inner qualities separate from external situations signals an opportunity to compare and contrast the importance of each. Relationships, finances, and priorities should all be reassessed on the basis of the inner qualities they exhibit in comparison to the external pleasure or stress that they generate. Your focus on internal values will tend to overshadow your need for external manifestation; consequently, you will make adjustments in your pursuit of materialistic and external goals.

Relationships are always more important for the love you give and receive than the external trappings; however, this issue will become more evident during this year. You may be in love with a special person who treats you wonderfully, but refuses to consider marriage. If you were attracted to this person because of his or her inner qualities, it is important to appreciate your feelings for one

another, and not get hung up on the legalization process. During this year you will tend to assess many relationships you see or experience for the inner and outer qualities which they possess. You can always find others who have the external trappings you desire without genuine caring. (Example: a verbally abusive marriage.)

External, artificial indicators of affection are meaningless. It is the inner and sometimes less apparent qualities that are important. Your attention on this issue will make the incongruity more obvious to you. Part of this discernment process can involve frustration with personal involvements which do not conform to your external expectations of what a relationship or partner should be. Love does not necessarily occur in neat little storybook packages, and frustration results from close attention to external storybook details which cannot truly satisfy internal needs. Outer limitations stress the importance of inner beauty. Because third parties may not easily see the qualities that you appreciate in your loved one, they may criticize your choice. Their criticism will encourage you to further define the inner beauty that attracts you. Those relationships which possess no redeeming inner qualities will seem void and unfulfilling. They can be discarded regardless of the external advantages they supply.

Values must be more closely attuned to the individual's aesthetic appreciation and therefore may be less consistent with social preferences and standards. One can see unique beauty in what might appear ordinary to others. Your taste in material objects might depend on the emotional qualities associated with those objects. Price or status is unimportant.

Venus retrograde symbolizes an emphasis on the quality of life over the desire for financial gain. The money you earn should not be as important as the quality of your working environment and your satisfaction while on the job. Focusing on your inner need for contentment and fulfillment allows you to let go of a large salary, and search for an enjoyable employment position even though it may involve a pay cut. If you continue to work at a stressful job because of the pay, you may be very unhappy. Materialism will not bring you true happiness this year or any year. If you must remain in a stressful job, develop a strong belief in yourself and your abilities. Do not focus on your inability to perform at your best when conditions are at their worst. Financially, this is general-

ly a time to be conservative with funds. You cannot make go
decisions about the quality of your life if you are more concern
with your monetary situation.

Social contacts and involvement in social activities are not e
phasized while Venus is retrograde in the solar return. You ne
more time to focus inward and may withdraw from some or a
social functions because you welcome the time alone. You mig
no longer enjoy these functions, or you might realize that certa
social relationships are too difficult and unrewarding to maintain.
For those individuals who require counseling, this can be a time
isolation. Withdrawal is common with Venus retrograde, but ge
erally any time alone can be put to good use. You could want to be
alone with one special person.

MARS RETROGRADE

When Mars is retrograde in the solar return chart, the indivi
ual must work with the process of self-motivation. This is a goo
time to work on a long-term project, especially one involving the
need to repeatedly push yourself towards achievement. It is ve
unlikely that you will be motivated by others since personal goals
will seem more important than the conflicting goals of others. If
you cannot motivate yourself or direct your energy in a use
manner, you will feel listless and tired. This is an extremely useful
retrogradation for those who are goal-oriented; unfortunately, it
can be a very counterproductive placement for those who are not

The way you choose to handle anger is symbolized by Mars
retrograde. Usually, there is a desire to avoid confrontation and
conflict. You may be unable or unwilling to express anger ou
wardly. Furthermore, you might find it difficult to be openly ag
gressive or even assertive given the situations you are involved
with. If this is so, you could resort to passive-aggressive behavi
or manipulation if it seems impossible to deal with a present situ
tion on a rational level. For example, if you are taking care of a
cantankerous and senile relative, confrontation and rational di
cussion will not improve your relationship, but refusing to enga
in conflicts and stressing the humor of the situation may. You can
manipulate your way around the old coot with love and unde
standing in your heart. Having Mars retrograde in a solar retu
chart signals the need to reassess the appropriateness of anger and
conflict in certain situations where it may actually be totally us

less. It is not the answer to all situations and you can learn to use other tactics.

If you are involved in difficult circumstances, you may not defend yourself against the criticism of others. And in fact, you would see yourself as responsible to some extent for the situations you are involved in. This is a time when you are more apt to get in touch with the role you, yourself, play in creating stress. You could blame yourself and be very self-critical of your own behavior. Positively, we can look upon this year as a time when you are more apt to see self-defeating situations and take corrective action. It is fairly common to realize the existence of at least one self-defeating situation or personality pattern during the year.

Mars retrograde, at its worst manifestation, can have a self-destructive interpretation. It is possible that you will place yourself (through your own doing) in a situation that causes you difficulty or pain. You will have the ability to withdraw from the situation, but might choose to remain throughout the solar return year. This may sound like a horrible manifestation, but it is not necessarily so. A few examples may help to clarify the meaning of Mars retrograde. One client forgot to use birth control and became pregnant at an inopportune time. She was the cause of her situation, and she chose to completely rearrange her life and have the child. Another client refused to accept that her two best friends were lesbian lovers despite the evidence to the contrary. Her inability to admit this to herself caused her unnecessary anxiety and tension. In each of these instances, the issues and difficulties were self-imposed and controllable in one way or another. The house placement of Mars will relate to the self-defeating or self-destructive attitudes and may symbolize this negative behavior in relationships, career practices, financial responsibility, etc. If one concentrates on the issues, solutions can be found and there will be no need to remain in compromising situations.

Interceptions

Interceptions in the solar return generally indicate some form of limitation or restriction, especially if there are solar return planets contained in the interception. Restrictions also apply to intercepted houses in a double chart technique wherein the natal chart is placed on the outside of the solar return. The immediate reaction to the word limitation is usually negative. We think of frustration and unfulfilled dreams. But everything has a limit. We put fenc around our yard to keep intruders off our property, not to frustrat our sense of privacy. Restrictions and limits are necessary facets of our daily lives. We cannot fully experience Jupiter without a sen of Saturn. There are times when we must limit the amount of de we are willing to handle. This issue could be symbolized by Venus intercepted in the 8th house of the solar return chart. (See figu 3.)

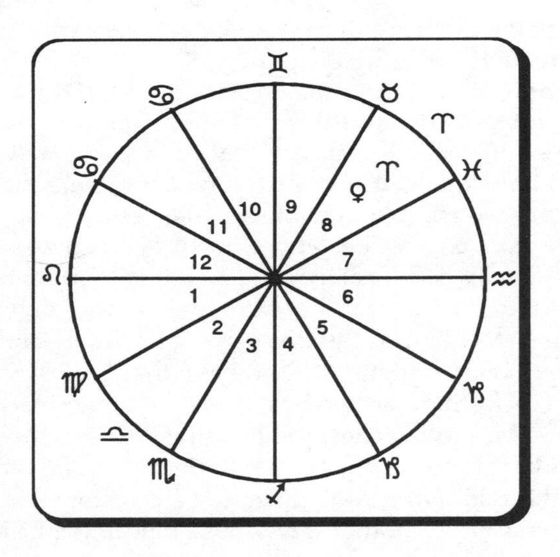

Figure 3

There are also times when our jobs are too demanding an begin to intrude on our personal time and energy. Feeling bur out, we may want to cut back to part-time employment. This w enable us to have more time to investigate other options or regen erate enthusiasm for our present or future position.

Interceptions signify limitations that we are willing to acce and may even welcome into our lives if they are a necessary con tinuation of our goals for the year. But interceptions are also indic ative of natural limitations which exist in everyone's life and mu be accepted and handled. Pregnancy and babies tend to slo mothers down (the 1st house of the solar return intercepted in the 5th house of the natal chart).

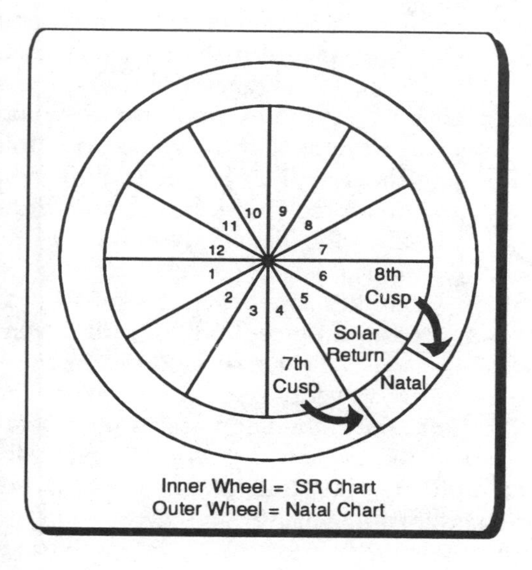

Figure 4

Lovers who honestly state that they are interested in an unbonded pair and not marriage are not likely to propose (the 7th natal house intercepted in the 5th house of the solar return). See figure 4. Within this context, interceptions do not imply either good or bad situations, only necessary limitations we need to incorporate into our lifestyle or realistic restrictions we need to recognize.

Hemispheric Emphasis

Presented below are the interpretations for hemispheric emphasis, quadrant emphasis, mode and element preponderances and lacks. Again, the reader should remember that narrow interpretations of singular astrological data are never the ideal. It is difficult to isolate one factor for interpretation without considering the rest of the chart. This is especially true of the following factors which tend to be general and particularly susceptible to contraindications in the rest of the solar return chart. It is important that you weigh the following factors according to their consistency or lack thereof with significant themes in the interpretation.

SOUTHERN HEMISPHERIC EMPHASIS

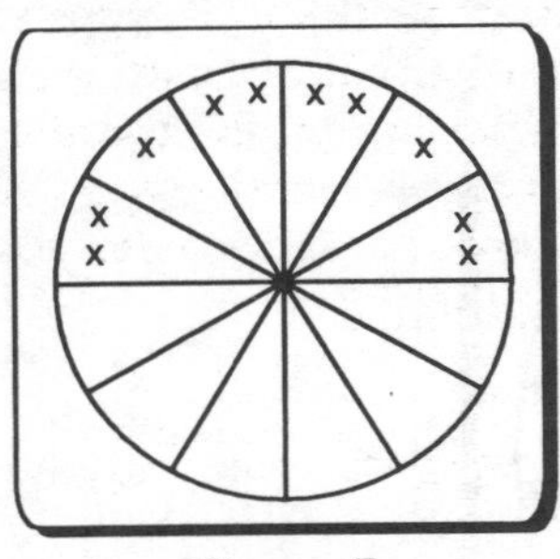

Figure 5

When nine or all of the planets in the solar return chart are above the horizon, this suggests that you are more interested in external manifestation and achievement than you are in internal processes. This is true even if the planets are mostly in the 8th or 12th houses. When the planets are in the 8th, the emphasis is on change. Granted these changes are brought about by an increased perception, but still the emphasis is on external modifications in lifestyle and the need to gain power over one's situation. When most of the planets are in the 12th house, the emphasis is on a year-long project which can lead to recognition in the following year. The project is generally a product of personal integration; however, the external manifestation is still noteworthy.

This is a time when you want to succeed in a tangible external way which is measurable and obvious to others. You evaluate your success by what others think of you and are less satisfied with subjective appreciation. Peer pressure influences you, and you can feel compelled to modify your personality, or behave in a certain manner. This is even true of 11th house placements where the two themes, one of individuality and the other of peer pressure, play off each other. What feelings and thoughts you do have may be controlled or lost in a flurry of activity as you strive for external achievement. The focus of attention is on what you can do and accomplish and not on how you feel about yourself internally. This is a good time to acquire objective information while testing your abilities to succeed in an external way.

Since you might devalue your own subjective thoughts and opinions this year, it is possible that you will seek advice from others rather than using your own head. You may even believe everything you are told, and can be deceived easily. Hollow goals built upon the expectations of others rather than personal fulfillment are possible if you lose touch with your own emotional roots and spiritual purpose. In extreme cases, this southern emphasis could denote the workaholic who sacrifices everything for success.

But if you consciously remember your roots, this can be a very productive time, one in which you use information and feelings to achieve goals that were formulated when many of the planets were in the northern hemisphere. This is a time for action, the counterpart to the planning stage. You may not need to focus inward at this time if you already have a solid inner base for the activity you are involved in now.

NORTHERN HEMISPHERIC EMPHASIS

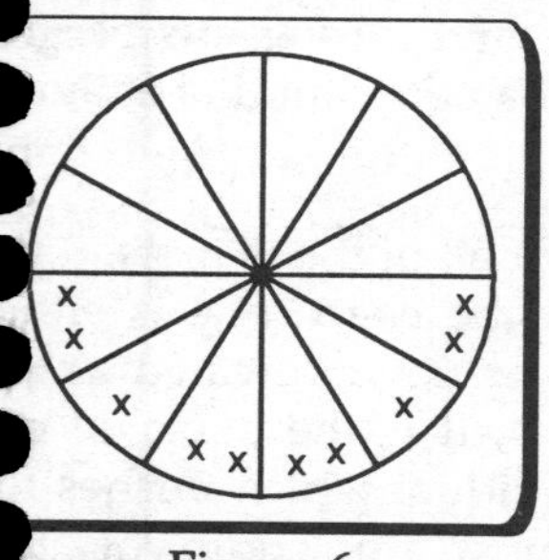

Figure 6

Charts with nine or all of the planets below the horizon indicate a period of internalization rather than a push for outward achievement. It is the internal integration which is most important at this time. This is true even of 5th house placements, wherein the inner personality must be made external through some creative or artistic mode of self-expression. It is also true of the 6th house, where health and work changes result from inner processes moving toward outer manifestation. Emotional satisfaction and fulfillment are the primary goals and motivating forces, therefore a lot of attention is given to feelings and subjective thoughts. What others think has little influence if the personal perspective is strong. Some will be withdrawn or somewhat reclusive during the year, using this time for quiet reflection and the fostering of inner strength. You might appreciate evenings at home with family and friends rather than social or business evenings out on the town. Use this time to become more rooted and in touch with yourself and others. Allow intuitive processes to grow so you will be better able to sense your true future direction.

The difficulty associated with the northern emphasis is that you may get so involved with your feelings that they overwhelm you and prevent you from living a normal life. You can neglect your responsibilities and lose all motivation towards success. It may become more important to express your negative feelings about your boss than to correct detrimental situations. If you cannot achieve a sense of fulfillment, you will grow bitter and sever relationships with little or no negotiation or warning. Nursing a wound is an inadequate substitute for emotional gratification.

This time can be used to get in touch with your sense of fulfillment and past achievement. You cannot understand where you are going if you don't understand where you have been. You cannot set goals for the future if you have not assimilated your past successes, and you can't branch out if you have not created a good support system among those you love. This is a time for building the foundation necessary for a new cycle of external achievement. Although you may actually be very successful during this period, you will still be laying the groundwork for a new round of activity.

EASTERN HEMISPHERIC EMPHASIS

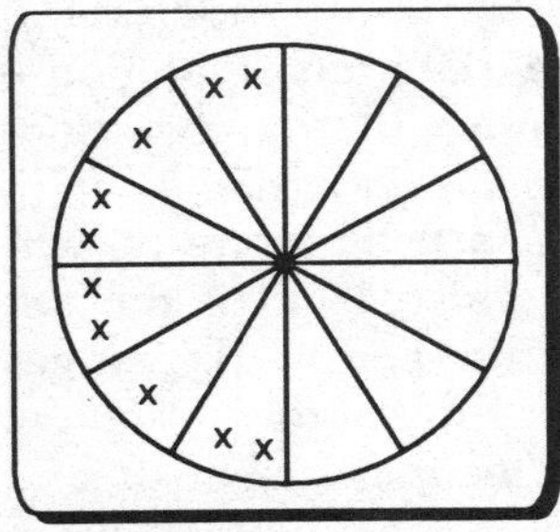

Figure 7

With nine or all of the planets in the eastern half of the chart, one is either self-motivated or not motivated at all. During the year, it is the personal desire of the individual which pushes for success in almost any endeavor. For this reason, persons with the eastern emphasis chart do not have to depend on anyone else for guidance or support. They are already inspired on an inner level. Use this year to develop self-motivation, self-direction and self-reliance. This does not mean that important others are missing from your life; but it does show a need for personal action and the realization that there are some things that others cannot do for you, and you must do for yourself.

This type of chart can indicate too much emphasis on self-interest without any regard for others. You can ignore the effect your behavior has on others, make impulsive decisions without consulting the significant people in your life, or refuse to compromise when there is a conflict. You may become increasingly comfortable with yourself while becoming dissociated from those whom you love.

This is a good time to feel your own power and sense of achievement on a personal level. You want to test your own ability to make things happen. Being successful this year can lead to further successes later in relationships and career. Many of your planets might be moving to the southern hemisphere next year, so this is a time for pilot studies and dry runs. You need to develop talents and self-confidence to prepare for future endeavors.

WESTERN HEMISPHERIC EMPHASIS

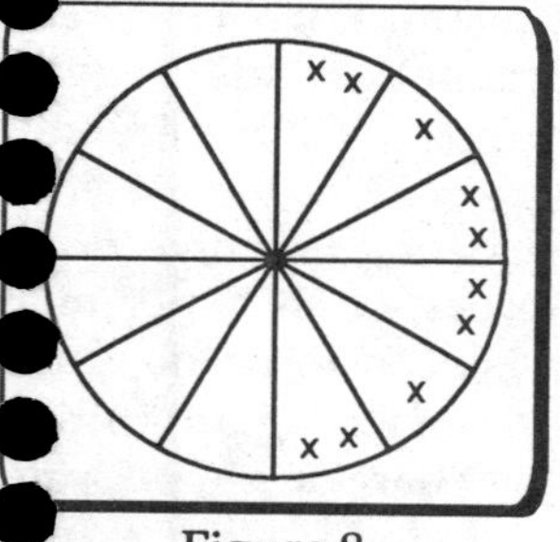

Figure 8

Nine or all of the planets in the western part of the solar return chart signify that actions are not based on independent needs since you respond in combination with others. Even those with numerous placements in the 5th house push for self-expression while maintaining relationships. Those with placements in the 9th house reassess their beliefs according to their interactions with others and their ability to manifest spiritual concepts in daily situations. This is a time when actions are defined with consideration towards others. Carrying this interpretation one step further, you may not want to act independently this year. Perhaps you will not initiate activities on your own, but prefer to follow others or become involved with their projects. Some individuals will be waiting for others to either make major decisions or lead the way into new experiences. During this time, you can grow more dependent on others or they become more dependent on you. The focus is on other people and what they can do for you, what you need to do for them, and what you can learn from each other.

The danger with the western chart is that you may become too focused on others and lose sight of your own individuality. You can spend the year catering to the needs of others while your own needs are not met. The strength of any relationship you are involved in is no guarantee that the relationship will be beneficial to your personally. You could stand to lose a lot: all your initiative, independence and self-interest.

But there are more positive interpretations. This is a good time to consult with others. You may achieve more through compromise and cooperation than independent action. Perhaps your goals can only be accomplished through the assistance of others and you need feedback and guidance to stay on the right track. Consulting with someone who is an authority in your field of interest or has experienced similar situations will make your tasks easier. This is a time for listening to others and gathering objective information.

Quadrant Emphasis

1ST QUADRANT EMPHASIS

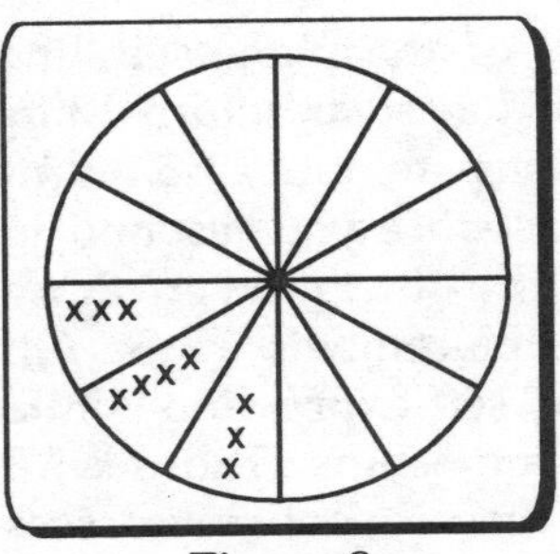

Figure 9

Nine or all of the planets in the 1st quadrant of a solar return chart can show self-preoccupation and a near-sighted personality. Your personal frame of reference may dominate your perceptions. Everything that goes on in your life is interpreted in a personal way. Very little is seen in an objective light unless there is a strong opposition somewhere in the chart or a natal propensity to do so. It could be very difficult for you to understand another's point of view or situation in an empathic way. It can also be hard for you to compromise on important issues. This is likely to be a very assertive year for you, and it is best to concentrate your energy on self-development and self-improvement. A 1st quadrant emphasis is commonly seen in those people who have given their all to relationships and family issues in the past, and now have time to work on themselves or build a life of their own.

2ND QUADRANT EMPHASIS

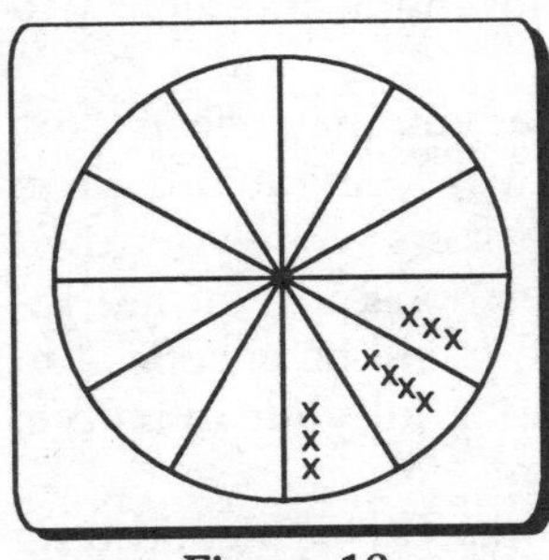

Figure 10

If most of your planets are in the 2nd quadrant of the solar return chart, there is an emphasis on expressing yourself both personally and creatively. This is the quadrant where self (the northern part of the chart) meets others (the western side of the chart). The emphasis is on achieving personality integration and individualization while still maintaining strong relationships with your family, lovers and co-workers. You may need to learn to express who you are in spite of the risk of disapproval and disagreement. Working creatively with self-expression and your relationships can help you achieve a complementary exchange of love and assistance at home or work, and in one-on-one exchanges. Creative abilities

will introduce changes in these areas, which could be beneficial for involved.

3RD QUADRANT EMPHASIS

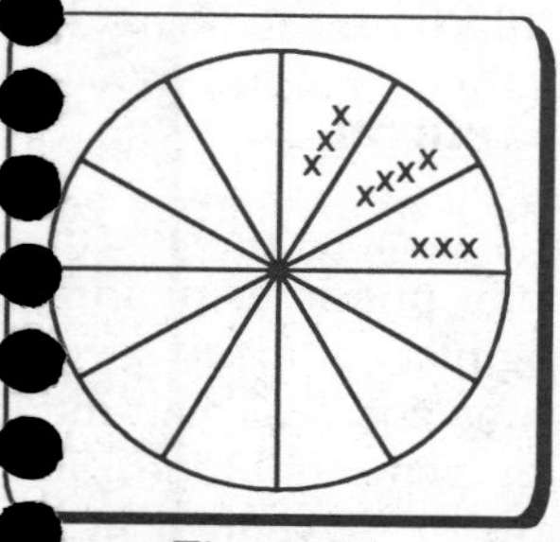

Figure 11

When most of the planets fall into the 3rd quadrant of the solar return, this indicates a strong involvement with others or one relationship in particular. If this is a love relationship, your focus of attention will be on the needs and wishes of this other person. You will tend to overemphasize his or her needs while neglecting your own personal needs and preferences. In the most negative sense, you can abdicate all power and play a passive role in events. If you are involved in a very difficult lawsuit, it is very likely that most of your energy and time will be drained away by the struggle. The year's activities may not be totally directed by you but hinge on the decisions and actions of others. This is a good time to gain insight from others, but it is not meant to stifle all self-interest. Focus on cooperation, not subjugation. Work with others, not against or for. Learn how and when to compromise, compete, negotiate and share.

4TH QUADRANT EMPHASIS

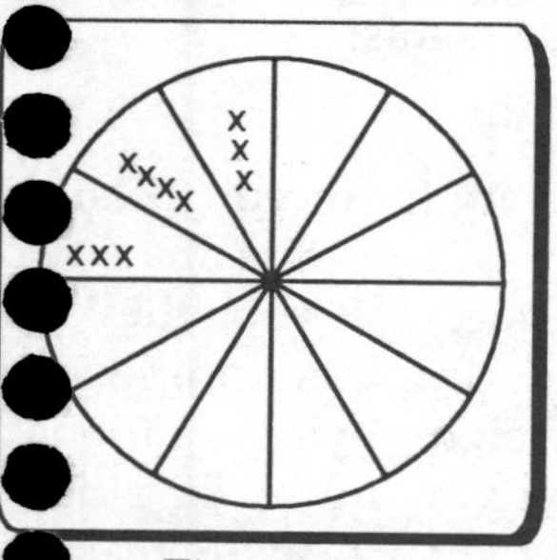

Figure 12

Heavy emphasis on the 4th quadrant of the solar return chart shows that a cooperative or uncooperative relationship with society dominates the year. Your personal form of integration into the social milieu defines which activities you will be involved in. You may choose to work alone and withdraw from social activities and associations entirely. Withdrawal can enable you to work more efficiently. Negatively, you might want to avoid working completely. Some will choose to function within a group environment, contributing to a combined effort. If you are interested in personal success and recognition, you could work towards being a leader rather than a co-worker or an equal. The goal is to find a position within or without the social environment which enables you to function in a positive way and contribute to society as a whole. Those who cannot may assume the role of social agitators or hermits. But most will define a place in the world and the process by which contributions can be made.

Modes and Elements in General

It is more difficult to assess the preponderance or lack of a mode or element than it is to interpret hemispheric or quadrant emphasis. This is because the slower moving planets spend several years in one sign and therefore have the same zodiacal placement in everyone's solar return. Even the faster planets (Moon, Sun, Venus and Mercury), are subject to patterns of movement in the solar return which restrict their placement in various elements. For these reasons, the house placement of any planet is more important than the sign it is in and the number of planets in any mode or element is subject to distortion. It might be artificially elevated or decreased. Therefore, when you look at the counts in any one mode or element, weigh these preponderances and lacks according to their confirmation of prevalent themes in the solar return chart interpretation. The preponderances and lacks are more likely to be important when configurations stress the interpretation. For example, the presence of five planets in cardinal signs is more noticeable in those who also have these planets in a T-square or Grand Cross configuration. Grand Trines emphasis the elements.

WELL BALANCED MODES AND ELEMENTS

When all of your elements and modes are balanced (meaning that there are no more than four planets and no less than two planets in any element or mode) this indicates that your thinking and actions are based on careful consideration of data from several sources of information. You are able to weigh physical, rational, emotional, and spiritual information when you are making decisions. You tend to be conservative rather than rash in making changes, even though you are open to change. At times you may be indecisive since you must take so much information into consideration; however, it is more likely that you will look at issues from all sides and not be overly influenced by any one personality factor.

FIVE OR MORE CARDINAL PLANETS

This can show a year with a tremendous amount of activity (which usually amounts to too much activity). You may expend your energy in several different directions by working on a number of projects. Each of these projects will be well-defined and equated with a personal need; however, you will feel torn between all the things you want or need to do. You are trying to do too much. If you overload yourself with too many activities, you will begin to procrastinate, miss deadlines and be forced into crisis management. Your nervous system will be taxed and you will begin to make impulsive decisions with little forethought. You need to be better organized. Perhaps you should finish one project before starting another, or you should seek assistance. If you must juggle a busy schedule, learn to streamline your routine and optimize your use of time and energy.

ZERO OR ONE CARDINAL PLANET

A lack of cardinal planets implies a need for simplification. It is common for the focus of attention to narrow from many activities to one major project or concern that dominates your perception for the year. For example, new mothers are usually primarily concerned with the care and feeding of the infant. Where many cardinal planets can show a scattering of energy, a lack of cardinal planets is more likely to be associated with a concentration of energy in one particular area.

However, the lack of cardinal planets can also imply a lack of activity. The pace of life can slow down because of external situations, or because of self-imposed limitations. You may decide that the schedule you keep is too hectic and consciously cut back on future obligations and activities. Occasionally illness will play a role, but generally this is only true if you are over-taxed and still have trouble saying "No" to new obligations. You will be less than energetic if you allow others to dictate your schedule and activities. Follow your own inner needs; this is a good time to work on the essentials only. Narrow your focus and push ahead full steam on that one project or issue that you want to concentrate on.

FIVE OR MORE FIXED PLANETS

When there are five or more fixed planets in the solar return chart, this is an indication that change or the lack of change is a major issue during the year. Changes are frequently viewed as all or nothing routines which can intimidate individuals into fostering approach-avoidance issues. One train of thought will emphasize stability and conservative action, while another stresses the fear of being rut-bound. Situations often appear to be either/or combinations with no options in between. Security becomes equated with familiarity even when what is familiar is difficult. A lack of stability is associated with making dramatic changes. In this sense, there is a tendency to accept things as they are and not look for alternatives. Fear of making decisions that would greatly alter your life could keep you immobilized and stuck in stalemated conditions. If you are faced with impending major changes but refuse to act on pressures and issues that are plaguing you, changes may be forced on you by external circumstances and these changes may be very dramatic.

Commonly a fixed preponderance indicates you are stuck in the same old situation again, particularly if you are involved in what seems to be a never-ending conflict or a deja vu relationship. The inability to look for alternative solutions to problems manifests as the inability to find new patterns of behavior. Stubbornness and a refusal to compromise are negative personality traits which can be particularly obstructive this year, especially if you refuse to see your own contribution to difficult situations. How easy it becomes to blame others! If only they would give in, give up or make changes for us, clearing the way for our own wants, needs, and ways of doing things. This is an unrealistic expectation. Positively, this is a time to draw on the the creative insights associated with the fixed signs, Taurus, Leo, Scorpio, and Aquarius, to develop new alternatives which optimize corrective action and change while minimizing instability. There are alternatives to either/or uncompromising situations for those who are willing to be creative.

If your life is already stabilized, then this is a time for building and maintaining resources. Further acquisitions can be had through creative use of what is already owned or known. It is an excellent time for finishing major projects that require plodding and sustained effort. But if major changes are occurring, this is also a good time to maintain a sense of stability and peace in the midst of disruption.

ZERO OR ONE FIXED PLANET

When fixed planets are lacking in the solar return chart, you may feel the need to initiate major changes and take greater risks. You will tend to welcome changes that come your way and actually see change as the natural way to improve your life situation. You have a desire for new experiences and can make major commitments to previously untried lifestyles. You might move a great distance, have a first child, get married, or experience a similar major transition. An element of insecurity or a lack of attention to security needs is associated with this modal lack. There is a tendency to jump now and think later. Your resistance to change may be so low that you lack stability during this time and make some changes unnecessarily. Use this year to be open to new possibilities. Prepare for times of great change by planning ahead.

FIVE OR MORE MUTABLE PLANETS

The more planets you have in the mutable signs, the more likely you are to be involved in changing circumstances. This is a year when you are better able to recognize subtle fluctuations in your situation as they occur. You are also aware of major changes as they occur little by little. Major transitions are anticipated long before they occur; consequently, you are able to adapt and prepare. Think of this as ad-libbing life. New developments and information necessitate constant reassessments and adjustments to your situation; therefore, this may not be the best time to come up with a master plan of how things should proceed. It will be much easier to handle daily issues as they arise, keeping a future goal or direction in mind. The key to handling changes at this time may be learning to ride the waves and make the adjustments necessary to stay on top of the situation.

There is a greater tendency to accept and adapt to changes as they occur than to initiate new changes yourself. Many times changes that other people are making or have made in the recent past directly affect your life. In some situations, you may feel that you have little freedom of choice and must adapt to surroundings that are different from those you have been used to in the past. At times, changes may be severe enough to cause anxiety, nervous upsets and stress, and you should remember to take time for relaxation techniques.

ZERO OR ONE MUTABLE PLANET

The lack of mutable planets implies that you are less willing to adapt to other people or situations. You have a strong desire to be yourself and refuse to modify your personality or change your plans for others. Instead, you may expect others to adapt to your needs and idiosyncrasies. If there are conflicts, you feel others should be the ones to compromise or take corrective action. This is probably not realistic or fair to all involved. More positively, you will be able to focus on a goal or task with a single-mindedness that will help you see things to completion.

FIVE OR MORE FIRE PLANETS

There is a great similarity between the solar return with five or more cardinal planets and the one with five or more fire planets. Both charts can show a lot of activity, but there is an essential difference. Those people with cardinal solar returns and those with fire solar returns wear themselves down in different ways. Cardinal activity people tend to scatter their energy in several different directions, whether purposefully or not. They can burn themselves out trying to do too many things at once. Fire people will tend to concentrate all of their energy on one or two projects only, excluding everything else, but they can still proceed to burn themselves out through their single-minded intensity. The fire personality is like a kid with a new toy, obsessed with one project, one job, one idea, one goal, one relationship, or one particular issue. But the fire can burn too furiously and be quickly spent. The cardinal person does a little bit of everything at once and the fire person does one thing furiously for a period of time. The danger is that neither will be productive. But the option for great achievement and productivity is there when a sense of timing is coupled with enthusiasm and inspiration.

You may need to compensate for heavy activity with quiet periods and withdrawal. Realize that you tend to exert energy in an cyclic manner and allow for slow periods. Optimize times of great activity by channeling your energy into constructive pursuits.

Intellectual spirituality and idealism are sometimes associated with a lot of fire planets in the solar return chart. There can be a love for the ideal or philosophical, but it is more common to simply feel inspired by future possibilities. This can be a time of great enthusiasm if you recognize your ability to create something new.

ZERO OR ONE FIRE PLANET

A lack of fire can signal the need to work with your present possibilities and immediate issues. You may not have time for long-range plans because everyday needs must be met. For example, if you have just entered the armed services, you will be involved with training for most of the year. Where your skills will take you once you are discharged will not be a concern. Now is the time to deal realistically with mundane issues needing your attention. Focus on the here and now rather than tomorrow. It is difficult to be inspired and creative with your life when your basic needs are so important. This is a good time to handle those basics and form a firm foundation.

You may not feel inspired. For some individuals, this is a year of disillusionment, lost enthusiasm and depletion of energy. If your philosophy is not consistent with your experience or is impractical, you must give up false hopes. If you have strong philosophical beliefs, test them in real life situations to see their practical application. If you have plans for the future, stop dreaming and start working. Get out of your head and into action. If you have no dream for the future, use this time to get your feet on the ground, and things in order. Your last dream may have been unrealistic, and now you are in a time meant for gaining experience and knowledge. Working with what is feasible will give you the sense of stability needed before you choose a new direction for the future.

FIVE OR MORE EARTH PLANETS

When there are five or more earth planets in a solar return chart, basic material needs are emphasized. This is usually a year when security and stability are sought and may or may not be attained. All kinds of work and employment are emphasized, even housework and chores. Employment-related changes and monetary concerns are common. Everyday survival techniques for the modern person will be stressed, i.e., how to get and maintain a good job, buy a car, pay your bills, or rent an apartment. Common sense is stressed as an important part of practical decision-making. Your ability to accomplish tasks and provide for basic material needs may be your most important goal during the year.

ZERO OR ONE EARTH PLANET

While the preponderance of earth is generally associated with attention to financial matters, the lack of earth is more closely associated with financial limitations. Implied is a lack of money, through a loss of income, mismanagement, or self-imposed limitations. If you usually have money problems, then the lack of earth in the solar return chart can indicate increasing or continuing economic difficulties during the coming year. You may find that you have reached your credit limit and can no longer spend money without planning. You will not necessarily have employment problems also, though sometimes that is the case. The lack of earth in the solar return chart suggests a lack of attention to financial concerns which can eventually lead to significant debt and overextension.

On the other hand, this is an excellent time to budget. Those individuals who are able to set their own limits use this time to become artificially poor and live on less money than earned. The lack of earth can indicate the desire to impose a strict system for saving money towards a large purchase such as a down payment on a home. In this case, the monetary pinch comes from your own savings program and not a lack of real funds or impinging debts. You can channel your money into various projects and expenses, leaving less to play with. If you are very concerned with your future financial security, vow to do something about it now.

For some individuals, the lack of money will not be a problem. You can place a greater emphasis on intangibles and therefore money is not a top priority. The quality of life is more important than the current financial state. If you are unhappy with your job, this is a good time to quit since career satisfaction will be more important to you than the salary you are paid. You are capable of living on less.

FIVE OR MORE AIR PLANETS

Years with five or more air planets are generally marked by an emphasis on rational thought and a lack of emotional expression. But there is the real danger that emotions are suppressed, and this possibility might be indicated by other aspects and placements in the solar return chart. On the other hand, perhaps emotions are just not a major concern. This is a good time for organizing and planning future events. Information can be acquired through greater objectivity. You will not take life too seriously, and can learn from everything that goes on around you.

Having a large number of air planets in your solar return chart generally says more about your mental state than your mental ability. The emphasis is on rational analysis rather than genius traits. Sometimes the emphasis on air is associated with learning and attending school, but this is not the usual pattern. There is not an overwhelming love of learning or a thirst for new knowledge as one might suspect; in fact, these characteristics are more closely related to the lack of air. An emphasis on air indicates learning tends to be tedious, repetitive, or very concentrated. If you are in school, you may be studying only what you have to in order to graduate, or working on original and exacting research.

ZERO OR ONE AIR PLANET

The lack of air in a solar return chart is associated with several different manifestations, but basically only one underlying personality trait. It implies that rational thinking is not the major component of your decision-making process during this year. Depending on what other element is emphasized, you may be very practical (earth), emotional (water), or inspired (fire). Because of the lack of air, you might not be objective. A personal perspective will predominate. Without the lightness of the air quality, you might take life too seriously.

Charts with little or no air can also indicate a year of little forethought. You may be impulsive, jumping first, thinking later. You respond to external events in a reactive way rather than planning your moves, especially if the water element is prominent. These reactions may be unconscious knee-jerk reflexes rather than considered responses formulated after a clear perception and assessment of the situation.

This lack can indicate inexperience if you are involved in a new activity or unfamiliar situation. Hence, it follows that you do not have the experience or knowledge necessary for what you are trying to accomplish. For example, suppose you buy an older home that needs a lot of renovation and you've never done this type of work before. You will undoubtedly spend many hours reading books and consulting with experts as you renovate the house. You learn by trial and error. Sometimes you do things well the first time and sometimes you do them over. For this reason, you may feel intellectually incompetent. This is not meant to imply that you cannot be successful in your endeavors this year. You can be very successful, but usually this will occur through a process of trial and error.

It is also common to feel uninformed. In its worst manifestation, it is a feeling of stupidity. But in its best manifestation, it is a thirst for knowledge and the courage to take risks attempting new tasks. This is an excellent time to gather information about new fields of interest, or to attend school.

FIVE OR MORE WATER PLANETS

The emphasis on water in your solar return is related to an increase in unconscious impulses and psychological factors. You may be more psychic or intuitive, but some negative psychological traits such as fears, phobias, dependencies, obsessions and compulsions may pop into your behavioral repertoire. These need not cause you discomfort, but you will have a tendency to react to life from a feeling level rather than from a clear avenue of conscious thought. This is a good time to work with your feelings and create a stronger bond of nurturing and caring with others.

You are apt to be more sensitive to emotional stimuli this year. A very positive manifestation would be a love relationship. Your feelings for family, parents, children and friends may be more poignant. But a preponderance of water can also symbolize an overwhelming emotional nature. You can lead with your heart, refuse to see things logically, and ignore your own best advice. There is a tendency to be more anxious or worried. If you feel overwhelmed by your own emotions and the problems of others, this is an excellent time to see a counselor.

ZERO OR ONE WATER PLANET

People with a lack of water in their solar return charts may find it difficult to express emotions or respond to emotional situations in a natural way. It could be more difficult for these people to make a complete emotional circuit, one in which they are nurtured, and in turn, are able to nurture others. Some of these individuals are victims of the strong-woman or strong-man syndrome. They pride themselves on being emotionally self-sufficient and self-contained, but may be experiencing a new depth of emotion which they need time to learn how to express more openly. There is a tendency not to do so.

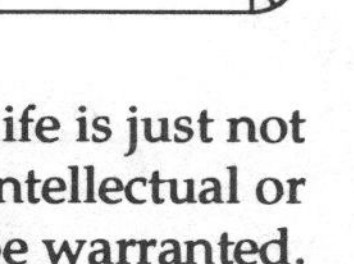

The third alternative is that the emotional side of life is just not a major consideration this year. This can be a time of intellectual or practical pursuits. A more rational orientation might be warranted. The danger is that the lack of emotional expression might be associated with sadness, loneliness or psychological stress, but this need not be the case. Other facets of the personality can take precedence.

The Ascending Sign

Generally speaking, the ascending sign in the solar return chart is read the same way as the ascending sign in the natal chart. The personality characteristics innate to the zodiacal placement will come through in the solar return. For example, those individuals with Taurus rising in their solar return chart may tend to be more stubborn and persistent during the coming year. Cancerian Ascendants may be more sensitive and emotional.

Those activities ruled by the natural house of the ascending sign can also be prominent. Following through with the same two examples, those individuals with a Taurean solar return Ascendant might consider their financial situation an important factor in the direction of the year's activities. Cancerian Ascendants might indicate a preoccupation with family considerations.

Because the meaning of the ascending sign is so closely related to the basic interpretation found in an introductory astrology book, the ascending signs will not be discussed here. The reader is referred to any text which explains the meaning of the rising signs and the basic activities of the corresponding houses.

Aspects in the Solar Return Chart

Aspects in the solar return chart can show psychological cohesiveness or a lack of integration. Use a ten-degree orb for conjunctions, sextiles, squares, trines and oppositions to the Sun and the Moon, and an eight-degree orb for major aspects between the other planets. These orbs are flexible and can be adjusted to accommodate configurations in the chart or your own personal preferences. Do not distinguish between applying and separating aspects. A transiting Saturn three degrees past a square to the Sun in a solar return chart still has a valid interpretation for the coming year. A planet having no major aspects to the other planets in the solar return chart symbolizes a psychological characteristic which is no well integrated into the psyche and lacks strong direction duri the coming year. The unaspected planet is more representative o time spent floundering than of time spent in pursuit of long-ter goals. Negative personality traits normally associated with th planet may disrupt planned activities. The major aspects (conjunc tion, sextile, square, trine and opposition) appear to have the stro gest association with solar return interpretation; however, th quincunx is also valid. It may be easier for you to understand and identify with the squares and oppositions in your chart than wi the softer aspects.

CONJUNCTIONS

Conjunctions indicate combined aspects of the individual psychological character. This combination can lead to a stronger focus on one particular issue, task or area of life. Several conjun tions can create single-minded purposefulness or misguided o sessions. For example, a woman with several planets conjunct the Sun in the 12th house was able to write a novel in the midst much commotion and many interruptions. She was able to tu out, and hence withdraw from the immediate situation. The major stumbling block seemed to be a tendency to dream about writing book rather than doing the actual work. Conjunctions can al show a combination of traits that are not easily blended, or which together become an overwhelming influence. A Pluto-Moon co junction in the 1st house of the solar return can symbolize a stro emotional nature that at first seems to overwhelm and dwarf ratio- nal thought. By focusing on fulfillment and long-denied or une pressed feelings, what first appears to be uncontrollable emoti gains meaning.

TRINES AND SEXTILES

Trines and sextiles symbolize compatible personality traits that are expressed easily and without inhibitions. Transitions can ev be made without a complete awareness of the process. One clie whose business was plagued by frequent moves adapted many o her services to mail-order practices so she no longer lost busine whenever she relocated. The change was very subtle, occurri almost without her awareness. The Sun was in the solar return 3rd house of communications sextiling her Moon in the 6th house daily work.

Trines and sextiles are not always beneficial. They can be associated with the uninterrupted continuation of negative behavioral patterns. They imply very little resistance; consequently, the individual is less likely to be aware of dangerous situations. For example, a male with five planets in the 12th house, numerous sextiles and trines and no hard aspects, grew more neurotic as the year wore on. He was unable to extricate himself from a relationship which affected his judgment and weakened his mental stability. He could not identify with the emotional damage being done.

SQUARES

Squares are representative of internal tensions and generally reflect either/or conditions or contradictory personality traits. Sometime during the year, the individual is forced to make a decision which either redefines or reiterates his or her personal needs. As an example, Venus in the 10th house of a solar return chart symbolizes a need to be materially successful in the professional arena; however, an individual with this placement felt that her creativity and personal freedom were being restricted by her nine-to-five job. Mars and Saturn were conjunct in the 1st house, squaring the 10th house Venus. She was eventually fired from her job for bad-mouthing the company president, but she used her unemployment compensation to support herself while setting up her own business. She wanted the financial security of a stable paycheck, but also the freedom of self-employment. By redefining her working situation, both needs were eventually fulfilled. Other dilemmas may be more or less crucial. They may or may not work out as well, depending on the individual's ability to integrate conflicting needs. Squares are indicative of dilemmas which mandate a decision about basic preferences.

OPPOSITIONS

Oppositions are frequently indicative of psychological energy we disown and attribute to other individuals. We project onto others traits and feelings we are not interested in integrating. These projected feelings are generally negative (anger, limitation, manipulation, lack of dependability, etc.), but we are also capable of projecting desirable qualities in the form of idealization. We tend to see these aspects of our own personality as representative of opposing forces or people in our lives who block or complicate our goals.

A teenaged boy who wanted to go to college thought that he would never realize his goal since his parents refused to provide the financial resources he needed. He wanted to be able to depend on them for help, and was angry that they seemed undependable. His Sun was in the 4th house conjunct Mars and opposed to Jupiter in Aquarius in the 11th house. But he had not investigated the possibility of developing resources on his own in the form of loans, scholarships and work-study grants. Since this person was intelligent, it was very possible that he could supply some of the much-needed funding on his own by recognizing his ability to function independently instead of insisting on being a total dependent.

Oppositions can represent seesaw points of view, and it is easy to be pulled between two situations and needs. The student mentioned above wanted both to be totally dependent, and also in pursuit of his educational goals. When he finally sought financial aid, he was able to integrate all his needs. He acquired the money for college by depending on his own abilities and resources. If you can integrate the two sides of an opposition, the middle ground may be much more productive than either end alone.

CHAPTER TWO

THE SUN IN THE SOLAR RETURN CHART

Introduction

The Sun is the most significant planet in the solar return chart. Its position by house shows the most emphasized area of life during the coming year, and how and where you expend the greatest amount of energy. The house placement of the Sun is reflected in the interpretation of the other planets, and may be the motivating force behind their symbolism. For example, a 1st house solar return Sun will show an assertive and subjective approach to life. All activities will be motivated by self-interest and the need for self-development. Most likely this individual will work alone on projects. If the Sun is trine Saturn in the 10th house of the solar return chart, personal qualities developed by the individual can be used in the area of career advancement. The self-improvement urge suggested by the Sun in the 1st house is reflected in the Saturnian placement in the 10th house.

Another individual with the Sun in the solar return 7th house will not want to work alone, but wish to work cooperatively with others. If Saturn is also in the 10th house of this chart, this person might form business partnerships or cooperative relationships during the year. In this example, the Sun's placement in the solar return chart has helped to explain the interpretation of Saturn's position in the 10th and indicated some of the methods by which career goals may be attained.

Patterns of Movement

From year to year, the Sun passes through the solar return chart in a clockwise direction, moving through the houses of the same angularity, falling into every third house.

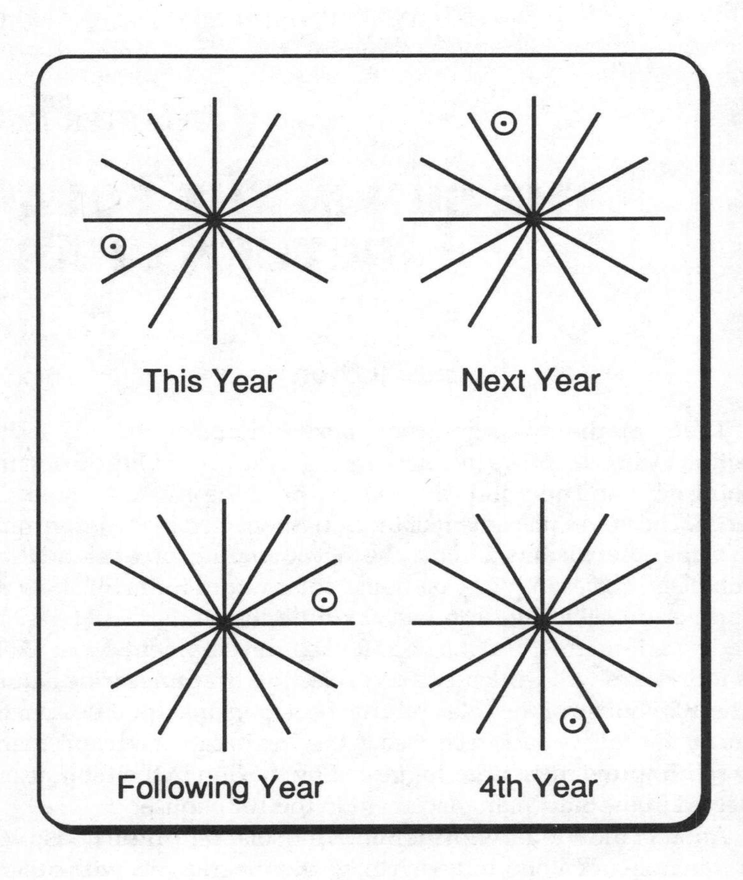

Figure 13

For example, the Sun in the 1st house in this year's solar return will probably move to the 10th house next year, assuming the individual remains in the same location. The following year it will move to the 7th house and the year after that to the 4th house. In this example, the Sun is moving through the angular houses. The Sun is just as likely to develop a pattern of rotation through the succedent and cadent houses. The Midheaven of the solar return chart jumps ahead three signs every year (plus or minus a few degrees), and this is the cause of the Sun's clockwise movement of three houses forward in the new solar return chart.

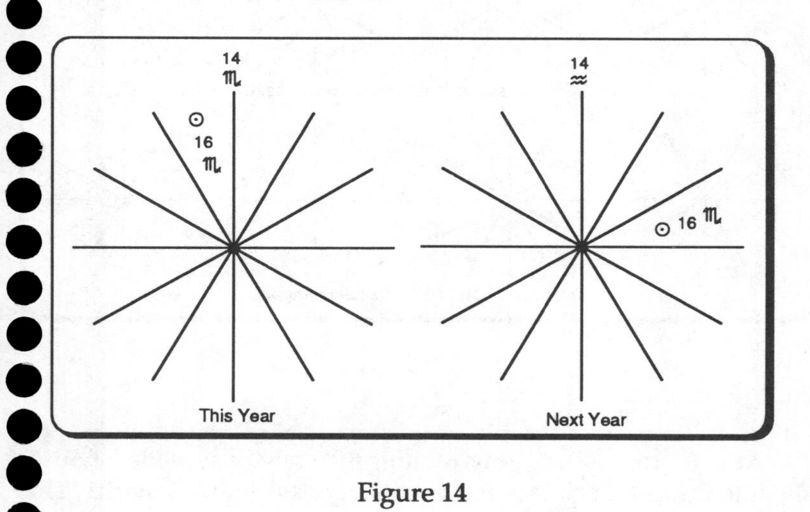

Figure 14

If your Sun is 16 degrees Scorpio and this year your solar return Midheaven is 14 degrees Scorpio, the Sun will be in the 10th house. Next year, if you remain in the same location, the Midheaven will be approximately 14 degrees Aquarius and your solar return Sun will probably fall into the 7th house.

The Sun rotates through a particular kind of house (i.e., angular, succedent or cadent) for approximately ten years. The reader should note that variations in this pattern will occur with signs of long or short ascension and with extreme northern or southern latitudes.

While the Sun is moving clockwise in the solar return chart from year to year, it also appears to be slipping counterclockwise. The pattern of movement through the angular houses gradually shifts to a pattern of rotation through the succedent houses.

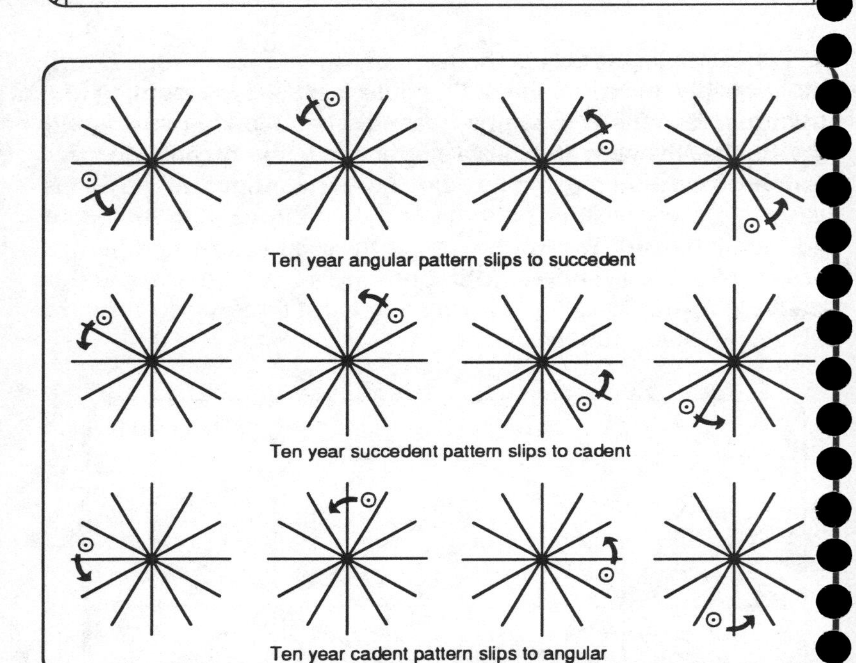

Figure 15

After ten more years, Suns rotating through the succedent hous slip into the cadent houses and eventually back into angularity. This pattern is dependent on the individual remaining in approximate the same location year to year. The whole process of clockwi rotation and counterclockwise slippage creates a cycle of 33 years. On the 33rd birthday, the new solar return angles closely approxima those of the natal chart for those natives who have remained in the birth location.

This 33-year solar return cycle is actually caused by variations the forward movement of the Midheaven. Imagine a clock with hand beginning to slip behind schedule. What was once 12:00 noon is now 11:55 am and later becomes 11:50 am. The clock continues to lose tin daily until eventually it has lost so much time (24 hours) that it is aga correct. Like the hands on the clock, the Midheaven of the solar return chart loses time from year to year.

In other words, the Midheaven of the solar return chart actually has two movements. It jumps ahead three signs every year (plus or minus a few degrees) and also loses a few degrees in the process. Since the Midheaven does not always appear to be slipping backwards, this secondary movement is sometimes difficult to detect.

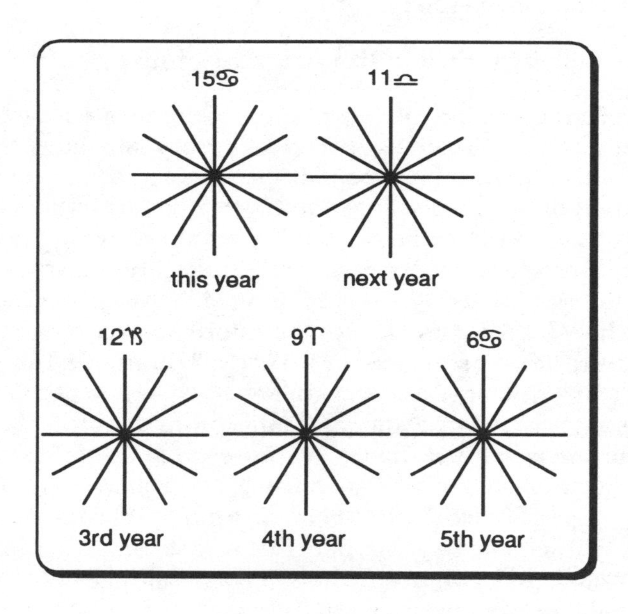

Figure 16

An example will help to make this clearer. Suppose your Midheaven is 15 degrees Cancer this year. Next year, it could be 11 degrees Libra, the following year 12 degrees Capricorn, then 9 degrees Aries. The fifth year, the Midheaven will return to the sign of Cancer, but it would have shifted backwards, perhaps as far as 6 degrees Cancer. As the Midheaven slips backwards in the zodiac, it causes the Sun to shift counterclockwise through the houses of the solar return. So the Sun's yearly movement can now be defined as jumping three houses clockwise while slipping counterclockwise very slowly from a rotation through the angular houses to the succedent and to the cadent houses.

The rotation through a particular type of house lasts approximately ten years. The starting point for rotation is shown in the natal chart. If we begin with an angular Sun, the Sun will rotate through the angular houses, then it will shift into the succedent houses for ten years and then eventually, it will shift into the cadent houses for ten years. This rotation creates a rhythm of initiation, stabilization and reinvestigation in the solar return cycle.

The Sun in the Angular Houses

The solar return Sun rotating through the angular houses focuses on initiating new situations or correcting old situations through various kinds of relationships. All the angular houses carry the connotation of new beginnings. In the solar return cycle, the Sun begins to rotate through the angular houses after having been in the cadent houses where the emphasis was on collecting information and reassessing conditions. When the Sun finally crosses over into the angular houses, it is time to use the information and knowledge gained to create new situations or make corrections in existing ones. Relationships with oneself, family members, spouses (partners) and business associates play a key role in this process.

If the Sun is in the 1st house, you are on your own. This is a time to relate to yourself and your own needs and abilities. You may be forced to depend on yourself to get things done. This does not mean that you will actually be alone, only that no one can do for you what you must now do for yourself. This is a good time to contemplate or begin a new career project or gain the knowledge necessary for a career push. Next year the Sun will probably move to the 10th house and you should get ready for an emphasis on work and a need for recognition that will come during that year. In this sense, the Sun's position in the 1st house of the solar return chart supports the Sun's coming placement in the 10th house. The Sun in either the 1st house or the 10th is a good position for assertive or aggressive action when it is used positively. These two placements usually follow one another and can work to your advantage if you use this time wisely.

The Sun in the 4th house is committed to emotional fulfillment. This house is the root or support system for the rest of the chart. Individuals with the Sun in this house are generally assertive when their emotional needs are not being met. They tend to ignore many outside activities in order to focus their attention on family matters

and emotional fulfillment. Relationships with family members are reassessed or duplicated with current friends and lovers. Emotional support established during this year is essential to personal development and career achievement during the following two years. The physical home or office environment may need renovation in order to meet future career needs. Changes that are made this year strengthen the support system necessary for future success and recognition. The Sun in this house will most likely move into the 1st house next year and the 10th house the following year.

Persons with **the Sun in the 7th house** are more interested in making cooperative relationships work for them than initiating projects on their own. This is a time to accomplish through the help of others, or in competition with others. On a more personal level, strong relationships are the focus of their attention. There is an openness to compromise and a greater tendency to seek advice from others, especially authorities in their field. Negatively, there can be a tendency to surrender power to a spouse, business partner or lover. The individual with the 7th house solar return Sun probably spent the last two years stressing personal development and career success (when the Sun was in the 1st and 10th house respectively). Now is the time to focus on relationships. Bonds that are strengthened during this year can grow to provide the emotional fulfillment and contentment needed when the Sun moves on to the 4th house.

The 10th house is a very assertive position for the solar return Sun. A 1st house Sun is also indicative of an assertive personality, but the range of effectiveness and desire is limited to personal matters. The Sun in the 10th house, however, can show an interest in career achievement and public recognition. Changes are not limited to the personal level, but manifest in a broader sense and with greater intensity and effect. Projects begun when the Sun was in the 1st house become a reality when the Sun is in the 10th. Those individuals who have not prepared themselves sufficiently for the push towards greater achievement and recognition may feel frustrated with their lack of success during this year. Each year is the culmination of the work done in the previous year. Self-development and the discovery of personal abilities and limitations should take place when the Sun is in the 1st house. Failing to work with this task could cause problems when the Sun moves to the 10th house. Tasks associated with the 10th house can support 7th house issues. Managerial techniques and successful business relationships can help you in the area of personal relationships when the Sun moves to the 7th house.

It is interesting to note that the Aries Sun, which is in a sign of short ascension, will spend little time in the 1st house of the solar return. Libra Suns, because Libra is a sign of long ascension, will spend a lot more time in the 1st house learning the lesson of self-reliance and self-initiation. Those individuals with an Aries Sun will spend a greater amount of time learning cooperation as the solar return Sun is detained in the 7th house.

The Sun in the Succedent Houses

When the Sun is in the succedent houses, the issues focus on the possible disruption of stabilized conditions, or a return to stability after a period of turmoil. A solar return Sun in a succedent house is like a person standing up in a canoe. There is the strong likelihood that the boat will rock in the near future or has recently rocked in the not-too-distant past. These years are times of transition. For those who feel rut-bound, this can be a time of exciting change. For those who have jeopardized their security in the recent past, this can be a return to greater stability. Boat-rocking themes focus on issues involving overindulgence, sex, money, values and self-worth.

Of the four succedent houses (the 2nd, 5th, 8th and 11th), **the 2nd house** seems to be the one most closely associated with a return to stability and traditional values. Change is not a characteristic of this house except in those instances where change is necessary to achieve greater security. The tendency is towards calmness. For example, it is common for the Sun to move into the 2nd house after a year in the solar return 5th house. Married persons who had an affair while the Sun was in the solar return 5th house tend to immediately drop the relationship when the Sun moves into the 2nd house the following year. The reasons given reflect a need for marital stability and a greater appreciation of the spouse's innate qualities. The 5th house is also associated with a strong need to express one's personality in a new and more forceful manner. When the Sun moves on to the 2nd house, these new forms of expression become standard personality patterns that others readily accept as customary.

When the solar return **Sun is in the 5th house**, you are more apt to express your true nature. Who you are does not change, but how you express yourself does. It is very common to grow more assertive during the year. Your ability to express openly who you really are this year is directly related to your ability to develop a higher sense of self-worth in the following year. Blocking expression of your true nature can lead to low self-esteem. The changes in the way you express and assert yourself can either disrupt close relationships or help to stabilize situations. Expressing your true nature while maintaining close relationships is part of the task associated with this year.

The **8th house solar return Sun** is generally associated with disruption. It is common to experience major changes that affect every area of life. Typical changes include relocating to another state or country, quitting work, becoming a full-time student or leaving your parent's home to live alone in your first apartment. Changes that occur with the Sun in this house are anxiety-producing, but not necessarily difficult. They increase your ability to express yourself next year when the Sun moves into the 5th house. One stabilizing change that is related to the 8th house is the desire to pay off debts.

The 10th house is the house of laws and traditions laid down by society as a whole. The 11th house, normally ruled by Aquarius, questions whether or not these laws created for the masses should apply to the individual. Those people with the **solar return Sun in the 11th house** are involved with the "Why not?" syndrome. They question the validity of laws and traditions that they find restrictive and have not yet personally experienced as useful. One married woman with the solar return Sun in the 11th house questioned the need for total fidelity to her husband. The Sun in the 11th house is associated with weakening inhibitions and the increased likelihood that changes will occur during the coming year. Prohibitions against open relationships no longer made sense to her. For those individuals with strong Uranian tendencies and a propensity for change, the Sun in the 11th house signals a push for freedom and disruption of standard practices. For those with strong Saturnian characteristics, change is less likely to occur. Some individuals experience this placement as a mental exercise capable of creating tremendous anxiety, but little external disruption. Others see this as a time to open doors to new experiences.

The Sun in the Cadent Houses

The Sun rotating through the cadent houses emphasizes the mind's ability to gather information from various sources and process this information into cohesive knowledge. Each one of the cadent houses relates to a different type of information, but the four of them collectively represent a time of preparation, reassessment, and reorganization. The activities of these houses may linger behind the scenes and often they are used as stepping-stones toward a major achievement which may not occur until the Sun moves into an angular house or close to the angle on the cadent side. The time spent in the cadent houses will increase your ability to make good decisions if you evaluate your options and situations in a knowledgeable manner.

The **3rd house** rules factual and superficial information without an immediate and obvious purpose. The emphasis is on learning, writing, studying and thinking without necessarily applying. There is an openness to new knowledge and a willingness to consider new topics. The material gained during this year will be reorganized and internalized when the Sun moves into the 12th. Just coping with everyday matters can consume a lot of time and energy. This is the house of the mind itself, and mental stress and communication difficulties can be associated with the Sun in the 3rd. The mind will always be thinking with this placement, and for this reason, rather than spending your mental energy worrying and brooding, it is better to have something important and exciting to think about. This is why learning and writing are so vital to a positive experience.

The solar return **Sun in the 6th house** is concerned with using information in an organized and useful way. Facts by themselves are not as important as the practical application. Hence, there is an emphasis on both health and work. Health information is gathered through intellectual sources and also from the body through weakness, sickness and biological feedback. Information gained from physical sensitivity to the body's needs is integrated with intellectual data to formulate good health habits. Work situations also allow you to use factual information in a practical way. This is a good time to try new work methods or to streamline your office operations.

The 9th house is the house of beliefs. In many astrology books, this is the house of spiritual and religious information, but it actually rules much more than that. It rules any belief or misconception that guides or misguides your life. Prejudice, religious intolerance, chauvinism, and all other ingrained stubborn beliefs that control your life are seen in the 9th house. When the Sun is in the 9th house, many "higher" and not so high beliefs need to be tested and worked with.

Generally there is at least one major misconception recognized during the year, but it is also possible to realize the fulfillment of a strong belief in your own abilities.

The **Sun in the 12th house** indicates a good time to organize and assimilate information you have collected. Reflection, introspection and even withdrawal can help you to personalize information in a meaningful way. This is the house of unconscious experiences. Factual and practical data become integrated with emotional insights to form spiritual understanding. This assimilation process is essential to the individual's multidimensional growth. The lack of integration associated with this house can lead to inaccurate or contradictory perceptions. Answers to pressing questions will not come from others; answers lie within. You will find life easier if you put together what you already know rather than take in what you find difficult to relate to. This is also a time to connect with Universal concepts of a religious or spiritual nature.

Planets Aspecting the Sun

SUN-PLUTO ASPECTS

The nature of a planet aspecting the Sun in the solar return chart and the aspect between them color the interpretation of both planets. Pluto aspects to the Sun (which usually last for a number of consecutive years) indicate an increased awareness of power. Awareness can come through conflicts and struggle with yourself or others, but it is the awareness that is most significant. How people get, maintain and use power for personal goals or for the control of others is more important than the struggle itself. It is the awareness which gives you a greater ability to control your own life. Power usage in confrontations and everyday situations will become more obvious, but even subtle shifts in power will be evident to you as you learn to recognize psychological motivations and manipulations. For some individuals this is a time to study psychology; however, many will notice examples of obsessive, compulsive, phobic, or manipulative behaviors in themselves or others regardless of their educational background. Unconscious needs are intensified and life becomes more complicated. Ambition and self-control are the more positive manifestations. Learning to deal with life on a deeper level is the hallmark of the Plutonian consciousness.

SUN-NEPTUNE ASPECTS

Neptunian aspects to the Sun indicate that the native is growing more intuitive and more sensitive. This sensitivity will eventually lead to a greater compassion for other human beings and a better understanding of relationships. Individuals become less egotistical and more vulnerable during these years, since they are likely to be confronted with their own human frailty or that of someone close. Involvement with alcoholism, drug abuse, martyrdom, dependency situations, and savior-victim type relationships is the more negative manifestation of this aspect. For some individuals, being less egotistical results in an unstructured personality which lacks control and direction. More positive manifestations include helping those in need, becoming more intuitive, and growing less concerned with selfish interests.

SUN-URANUS ASPECTS

Uranus aspecting the Sun suggests that the individual desires to make changes, possibly in rapid succession. Generally, any solar return year that has a major Uranus-Sun aspect also has a corresponding significant life change or development such as a pregnancy or birth, career or job transfer, relocation, illness, etc. Changes tend to be more disruptive and less controllable when they involve a conjunction, square or opposition aspect, but all aspects can ultimately indicate beneficial changes. Issues involving boredom versus originality, or creativity and freedom versus restriction, are common.

SUN-SATURN ASPECTS

Sun-Saturn aspects in the solar return chart tend to imply a sense of structure. Whether this structure becomes supportive or restrictive is up to the individual's ability to handle Saturnian issues in a positive manner. This is not meant to be a depressing time, but it does entail stark realism. Accurate perceptions of existing situations are essential to either accepting or changing future expectations. The refusal to accept responsibility for one's own life situation or to work with obvious limitations can lead to frustration, isolation and loneliness. Limitations are not an essential characteristic of Saturnian aspects, but denote a need to be more realistic and patient. Changes are slow and involve careful planning, hard work, and discipline. Many times a major project is being worked on for most of the year.

SUN-JUPITER ASPECTS

Although Jupiter transits a new sign every year, Sun-Jupiter aspects do not occur in every solar return chart. The major task associated with Sun-Jupiter aspects is expansion of the personality into new areas of expertise. Hopefully, this expansion will be consistent with the individual's philosophical beliefs and spiritual goals. Jupiter's sign may be significant in some way, though usually it is the house placement that is important. For example, Jupiter in Capricorn can suggest a preoccupation with materialism. If Jupiter is also in the [illegible] house of the solar return chart, the individual may be looking to buy a bigger house in a more prestigious neighborhood. The main danger associated with this planet is a tendency towards excessive behavior and a refusal to curb personal needs and desires in consideration of others. Beneficial opportunities are associated with Jupiter; however, there is no guarantee of a positive return. You can augment the possibilities through enthusiasm.

SUN-MARS ASPECTS

Mars symbolizes the energy necessary for successful accomplishments. When Mars aspects the Sun, it is a good time to work on a project that requires a great deal of energy to complete. Success can come in the form of personal or professional achievement, or with defense. Relationships tend to be competitive, and you need to balance self-centered drives with the needs of others. Learning to deal with aggression, conflict and anger in an effective rather than detrimental way can enable the individual to handle negative situations positively. Less spiritual themes involve destructive or self-destructive urges. Prolonged anger creates blockages which waste energy in a cycle of negative emotions that have no real purpose or goal. It is better to direct efforts into positive endeavors.

SUN-VENUS ASPECTS

Just as in the natal chart, the solar return Sun is never more than [illegible] degrees from Venus, so it does not form any major aspects other than the conjunction. Venus conjunct the Sun indicates that personal reward or sacrifice in the form of money, self-esteem, values, or relationships is intimately tied up with the goals of the Sun. Laziness might be very detrimental, while one can reap great benefits from a personal best.

SUN-MERCURY ASPECTS

Just as in the natal chart, the solar return Sun is never more than 28 degrees from Mercury so does not form any major aspects other than the conjunction. Mercury conjunct the Sun indicates that the intellectual mind is actively involved in the goals of the Sun. Reading, writing, learning and communication in all forms will be important to the pursuit.

SUN-MOON ASPECTS

Sun-Moon aspects indicate either compatible external and internal goals, or a lack of agreement between the conscious and unconscious levels. Depending on the aspects and the individual's ability to integrate diverse needs, this can be a time of harmony or conflict. Common themes involve domestic and career needs. These two areas of life will complement each other or divide the individual's attention in a stressful way. Negatively, career moves may disrupt the family. Reputations thwart ambitions. Domestic responsibilities disrupt your work schedule. More positively, changes in the home coincide with professional moves. As children go off to college or school, parents are advanced to new positions. Relocations are welcomed by all family members. The unconscious need for emotional fulfillment is played out consciously in the external environment.

The Sun in the Solar Return Houses

THE SUN IN THE 1ST HOUSE

The Sun in the 1st house asks, "What can you accomplish on your own?" This is a time to do something by yourself and for yourself. It is likely that you have a project in mind that only you can complete. The 1st house is the house of self-discovery and you need to explore your talents and limitations in order to assess the feasibility of the project you have in mind. In reality, the project is only a focal point of attention urging you to investigate your own abilities. The true power of this placement lies in what you discover about yourself. Use the newfound traits for personal accomplishment. This is an excellent time for self-improvement, self-motivation, and personal change. Remember to initiate matters on your own since no one is going to push you along. Don't wait for others to do things for you. This is not to say that you will live alone, but this is a time to test your own strength and self-sufficiency. You want to be the one to make things happen.

This year is actually achievement-oriented, but all achievements are personal successes and not career triumphs. Advances are not so apparent to others since they tend to be internal rather than external. Growth is measured by the satisfaction that comes from reaching personal goals and making personal changes. During the year, character traits which either help or hinder your ambitions will surface. It is your job to foster the more positive qualities and to overcome deficiencies. This can be a year for healing and recuperation from a difficult relationship, or it can become more. It can be a time of great personal progress; it depends on what you are ready for.

If you start something career-wise this year, it will tend to grow and blossom next year when you will receive wider recognition for your work. Use this time to prepare for the career push that could occur next year when the solar return Sun will most likely be in the 12th house. You may need more professional training, experience, or education. But even personal changes and improvements you make now can support career or business goals later.

During the year, you tend to be oblivious to the needs of others unless there are strong oppositions in the chart or other indications. With a strong 1st house emphasis, the solar return shows a lot of subjectivity and diminished objectivity. Your perception of life narrows as you focus on your needs, abilities, strengths and weaknesses. This shift in emphasis is probably essential for the task at hand. You may need to keep your attention highly focused in order to channel your most positive traits into highly creative and productive tasks. Communicating your enthusiasm to others will help them to understand the importance of your endeavors.

THE SUN IN THE 2ND HOUSE

In general, the Sun in the 2nd house is not a sign of monetary gain; in fact, it is more likely that you will feel underpaid. Check Venus, her aspects, and the other planets in the 2nd house to understand the financial picture. This is the year to reassess your worth as an employee or business owner and decide how much you should be earning. You may find that your work is not fully appreciated and it is really worth more than you are presently being paid. This is a time to map out a plan to earn more money; however, your salary will probably not increase much this year. Usually, individuals with 2nd

house Suns do not get a pay raise until the following year when th
Suns move into the 11th house, the 2nd (money house) of the 10
(career).

If you are running a business or a household, learn to budget yo
money or work with an accounting program. You need to p
attention to the way money is handled and spent; it may be slipping
away wastefully. Financial practices that you incorporate this ye
can lead to money saved for future major purchases and projec
Lack of fiscal responsibility can lead to financial problems and
limitations.

Reassessing your sense of self-worth will also involve analyzi
how you are being treated by others. You may find that you do not
command enough respect. If you have very little self-esteem you m
be involved in physically or verbally abusive relationships, but it
more common to have established a pattern of devaluing your own
needs and abilities in comparison to the needs and abilities of othe
If others come first in your life and you always come last, it is time
make adjustments. Although you have probably helped to establish
this negative pattern, others have helped to reinforce it and no
everyone needs to change consciously. Now is the time to stand up
yourself. Expect equal consideration; if you must, demand it.

Learning to value yourself is just one side of this issue. All valu
need to be prioritized. What you once thought was desirable is n
no longer attractive. You may grow more materialistic; or conversely,
you may stress inner qualities rather than money. But in either ca
you will need to make decisions and set priorities during the year
moral or ethical conflict is common. Usually there is a shift towards
more traditional codes of behavior.

Some individuals make a definitive moral or ethical stateme
They feel compromised by their jobs, situations or relationships, and
feel compelled to stand up for what they believe in, ignoring the
physical, emotional or financial consequences their stand will bri
This is the placement of the whistle-blower.

The 2nd house is also concerned with overindulgence, such as
overeating, smoking, drinking, impulse spending and promiscui
You are more apt to be aware of excessive behavior during the year
and consequently seek to control it.

THE SUN IN THE 3RD HOUSE

The Sun in the 3rd house of the solar return emphasizes yo
intellectual abilities and mental stability or the lack thereof. Intell

tually, this is a time to gather information. Your mind is very active and you will want to read everything, know everything, and think about everything. Ideas abound and you are open to looking at life from a new perspective. But organizing your mind may be difficult, which is why this is a good time to study, write down your thoughts, or purchase a computer.

One way or another, your mind plays a crucial role in the events of this solar return year. The mental processes are key to the activities you are involved in, or central to the major problems you encounter. Your mind can work for or against you, and make or break the year. Acquiring knowledge may be the main focus of your attention and it is certainly advantageous to learn as much as you can during this time by attending school or taking a course. But you can feel intellectually inferior or frustrated by your educational attempts if you allow nervousness, indecision and impulsiveness to affect your ability to think clearly and logically.

Mental instability is a possibility with the Sun in the 3rd house of the solar return. The Sun here is an even stronger indication of depression than Saturn. Check Mercury and its aspects to better understand the Sun's manifestation in the solar return chart. Depression, anxiety, irrational thinking, confusion and neurosis are possible extremes. Your mind is working overtime, and if you do not direct your thinking to-ward meaningful or educational pursuits, mental difficulties can arise.

You are probably more interested in thoughts than feelings this year (depending on the position and strength of the Moon and Pluto in the solar return chart). The Sun in the 3rd signals an emphasis on the thinking processes, so you will spend more time thinking about feelings than actually feeling them. Feelings will be analyzed and dissected rather than felt as you try to understand them from an intellectual perspective. This is not to say that you are cold this year, but you will have a greater tendency to screen your feelings and make logical decisions rather than emotional ones. If the Moon and/or Pluto are prominent in the chart, you may find it difficult to integrate what you feel with what you think. If this is the case, unconscious material may compete with rational thoughts for control of your thinking processes. The Moon in the 1st house is a major contraindication to the dominance of thoughts over the emotional nature.

The Sun in the 3rd can also show community involvement or activism. It becomes important who your neighbors are and what your neighborhood is like. You can expect to be more involved with those around you. This can be on an individualized level (one neighbor needs your help or one neighbor creates a problem) or this can be on a community level.

All means of communication are stressed, and you might want to work on your communication skills to increase your effectiveness with the spoken or written word. Some individuals will even take a course in communication-related topics such as effective listening, resumé writing, advertising, etc. This is a time when you are more likely to use the telephone or the mail system regularly to communicate with people.

THE SUN IN THE 4TH HOUSE

The Sun in the 4th house is basically concerned with finding a physical, emotional and spiritual home or niche. There is a special place for everyone in the Universe, a place that reflects their spiritual purpose and fulfills their emotional/physical needs, a niche that defines, directs, and supports. It is the place where you function at your best while feeling protected and encouraged by your environment. On a physical level, this niche is a warm and comfortable home. On the emotional level, it is a supportive relationship, and spiritually, it is a fulfilling purpose or goal. With the Sun in the solar return 4th house, you need to find that niche on the physical, emotional and spiritual levels. This is a time for grounding your existence with purpose and establishing the roots that will support you in the years ahead.

Finding the physical niche means finding or constructing a home that is a comfortable place of revitalization. Your pleasures are simple and center around whatever you call home, but it is not uncommon for "home" to become unsuitable during the solar return year. Your needs have changed and you must now make some adjustments in your surroundings. With only the Sun in the 4th house and not too many aspects, you probably only want to redecorate or reorganize your home situation. Those individuals with Uranus or Pluto also in the 4th house may want to make sweeping changes that tend to involve greater disruption.

The search for a home on the physical level leads many people to purchase new homes or fix up their old ones. They need to have a physical place of retreat in order to replenish their energy. The physical niche is crucial since it is the beginning base of operation which goes on to support both the emotional and spiritual levels. The external emphasis on purchasing or repairing a home reflects the internal emotional need for a supportive environment. If you examine your living or working environment now, you will discover that there is something which physically impairs your future goals. In most instances, there is a need for a private place for thinking, working or studying. This need for space is important and must be satisfied if you are to build a multi-level supportive environment which will augment your future abilities. Work begins with the physical home. Watch the symbolism of what you are doing to the physical home. As you are satisfying physical external needs, correlations can be drawn to improvements occurring on the emotional and spiritual levels.

The Sun in the 4th house is also focused on the emotional level. This is a very strong position for individuals seeking emotional satisfaction and fulfillment. For this reason, you will tend to be more vocal about your feelings of loneliness and lack of emotional support from friends, family, co-workers and bosses. You will also be more appreciative of the support these people give you. Loneliness may be the result of delayed grief. You may grieve for those who have died or left even though the separation occurred several years before. The Sun in the 4th is meant to be a time of emotional healing; it may not be a year for personal achievement, independence or assertiveness if you are more concerned about emotional fulfillment and interconnections with others. Gains can be made internally rather than externally.

Family involvement grows and this is a good time to improve or renew relationships with estranged family members. Resolve old resentments which clog up the emotional nature, making it more difficult to express positive feelings. This process of emotional housecleaning may involve reevaluating your childhood and your relationship with your mother and father. Your memories during this time may not be especially pleasant; working to give yourself the emotional support you will need in the future can involve reconciling past conflicts. If you still cannot get what you need emotionally from your relatives, then you will seek comfort elsewhere and grow apart.

Dependency issues are common. An increased need for physical or emotional support can evolve into an obvious dependency. You may be physically or emotionally dependent on someone else as a result of real or imagined limitations. Those individuals who are already emotionally unstable can regress and exhibit childlike characteristics. This is a very negative and, fortunately, very rare manifestation.

You may feel physically and emotionally abandoned by your parents, and their lack of support becomes an issue. They would feel that your demands are too great; you would say their help is nonexistent. On the other hand, you might be asked to give your support to others. Sometimes grown children or elderly parents turn to you for assistance if they are emotionally drained or physically weakened themselves and greatly in need of help. It is good to remember that we all have different needs at different times. They wax and wane according to our spiritual, emotional, and physical diet.

Usually the Sun in the 4th moves to the 1st house in the following solar return year, so now is a good time to practice using your emotional and intuitive processes in a positive way. They are important sources of information, essential to your growth. If you get in touch with your emotional-intuitive nature, you will learn to trust the information you receive. Gradually, as the year moves on, you will become aware of factual or physical evidence supporting your intuitive impressions. You need to break through emotional fears and begin to work positively with your insights so that your intuitive feelings will flow easily. Next year when your Sun moves to the 1st house, you will have a good information base for independent action and personal assertiveness.

And finally, you must find a spiritual niche in this world. Everyone is here for a purpose. Your spiritual purpose may be renewed, clarified, or changed at this time. You need to redefine what your higher purpose will be in the coming years, and how you can best serve the needs of the Universe. You will be guided in your search for what makes you feel fulfilled. This is why emotions are so important this year—they put you in touch with your spiritual purpose. Next year (when the Sun will probably be in the 1st house), and the following year (when the Sun will most likely be in the 10th house), you will be involved with expressing yourself more assertively to the world. With a clear sense of spiritual direction, a strong intuitive nature, and a supportive environment, confidence will be maintained and growth will occur easily. You will use the supportive physical environment, emotional connections and spiritual goals you establish this year to back up your efforts in those years to come.

THE SUN IN THE 5TH HOUSE

The Sun in the 5th house emphasizes the need to express yourself more fully. When the Sun is in the 1st house you try to find out who you are, but when the Sun is in the 5th house, you know who you are, and have a stronger-than-usual urge for expression. Having your Sun in this house can mean watching your personality bloom. This is a wonderful position for those who have been compromised in the past and now feel the need to be more assertive. The outer expression of your personality should become more useful to your future goals and also more consistent with your inner qualities. "To thine own self be true," should be the motto for this placement. But keep in mind also the desire to maintain relationships with others. Although self-expression may disrupt relationships somewhat, the purpose of this placement is to foster good relationships while increasing your self-expression.

Self-expression can take several forms, some of which are artistic and most of which are creative. Artistic pursuits include painting, writing, poetry, and music; however, you may concentrate on a more mundane project such as creating your own lesson plans, developing your own business style or founding your own nursery school. The medium for self-expression is not important, it is the self-expression itself which is creative, and it need not necessarily be artistic.

The Sun in the 5th house can also show a strong involvement with children; if you are a parent, they may be the major focus of your attention. If you have a tendency to express yourself through your children, you must be careful not to dominate their lives. On the other hand, this can be a rewarding time, one in which you are very proud of your children and their achievements.

There is always the possibility of romance with this placement, and if a relationship occurs it is usually very exciting. Sexual attractions play a major role in determining to whom you are drawn. Romantic interchanges and affairs are likely. This can be a heart-pounding infatuation at its best, but remember that the 5th house rules unbounded relationships rather than marriage. Although you may discuss marriage with your newfound love, it is very unlikely that you will tie the knot this year.

Because the 5th house is also the house of speculation, you are more likely to take risks this year. You do not necessarily gamble or speculate (though you may), but you are willing to bet on your own abilities. The chances you take may or may not involve money, but you are open to asserting yourself and trying new things. You readily risk failure because of a desire to stretch the boundaries of self-expression.

THE SUN IN THE 6TH HOUSE

Health is a concern when the Sun is in the 6th house, and you should start taking better care of yourself. Your health needs and/or attitudes are changing and it is time to create a line of communication between your mind and your body. Pay attention to what your body is trying to tell you; it is more important that you eat and exercise right according to your body than according to some nutrition book or philosophy. Health should come from the inside out. Pay close attention to the effect of various foods on your system and make adjustments accordingly. If you ignore your body's warnings during this year, you can expect to have new or continuing health problems. Usually major illnesses are only a problem with this position if you totally neglect or abuse your body.

The emphasis of the 6th house is more on the daily details of work than the overall career. Consequently, you are more job-oriented than career-oriented with this placement. The difference is very subtle. You are aware of major career issues when the Sun is in the 10th house, and you are more apt to work for promotion, recognition, and career change during this time. Important decisions regarding profession are possible while career fulfillment is likely. But when the Sun is in the 6th house, your profession will tend to be stable, and the emphasis will be on immediate job issues and working situations. It is the daily procedures which are in a state of transition. This is an excellent time to reevaluate your work routines and make adjustments towards greater efficiency. Even if you love your job, you will find parts of your day tedious and boring. Streamline your operation and make changes where appropriate. You may be asked to adapt to changes beyond your control. Mergers and reorganizations can occur, affecting your work environment, position or job description. In a very negative manifestation, you may be laid off or forced to work overtime. Personal problems, illnesses, or family matters can also disrupt your work schedule. Work schedules may be changed and flexible hours might be an option.

THE SUN IN THE 7TH HOUSE

The Sun in the 7th house indicates that relationships are more important to you than your own individual identity. You are more interested in relating to others (and possibly someone in particular) than pursuing personal goals or working alone. It is very likely that your Sun was in the solar return 1st house two years ago and in the 10th house last year. For these two years, you probably stressed independent action and self-sufficiency. This emphasis on self and self-motivated endeavors is over and now you have a strong need for companionship, sharing and feedback. Focus on relationships, and spend time with those you love who may have been feeling neglected. Their needs seem stronger than yours and it feels right to support those who have supported you over the past two years. It may not be easy for you to shift from a self-centered orientation into a partnership commitment, but you will find that fulfillment comes from sharing and cooperation. Your greatest successes occur when working with others or through the assistance of someone else. Generally, this is not the time to do things on your own.

Relationship problems are possible. Renewing old relationships can be difficult because others may have established a pattern of getting along without you if you were too distant in the past. Hopefully, they appreciate the time you are now able to spend with them. An old and useless relationship might end during this time so that a new and rewarding one can form. Use this time for a transition to a new and exciting commitment. Generally, new relationships are of a personal nature, but occasionally the emphasis is on a business partnership. The 7th house is the house of marriage, partnerships and strong relationships, and it is likely that you will be deeply involved in one of these types of relationships during the year.

Partnerships can be either supportive or restrictive. There isn't a guarantee that relationships will be beneficial to you, but you mold them with your responses. This is an excellent time to work with someone on a project. Consult those offering objective feedback or expert information in your field of interest. You can accomplish more through cooperation or competition than you would have accomplished on your own.

Some individuals experience this placement as restrictive, especially if they always emphasize the needs of others without considering their own needs. The Sun in the 7th house can indicate a very consuming relationship, one which totally compromises your individuality and personal needs. In this situation you become too passive, refusing to assert yourself at all, while allowing a relationship to become everything as you become nothing. You accept limitations passively, are easily imposed upon, and tend to follow, never lead. You grow more insignificant the longer you remain in this type of relationship. Most likely, you are trying to hold together an impossible partnership, one that should be ended. If you continue to bite your tongue so things will run smoothly, your own needs will not be met and the strain of meeting the needs of others will drain you.

Unless your partner is physically ill, your dedication to another should not be so demanding as to be all-consuming. Should you choose to remain in a detrimental relationship with demands this great, be aware that the situation is probably psychologically unhealthy for you and may eventually involve verbal or physical abuse. Learn the fine art of negotiation. This is not the time for complete surrender. One hand washes the other and it is through cooperation that you progress and grow. The task is to learn to compromise and share within a meaningful and fulfilling relationship. Give willingly without allowing yourself to be used.

This is the time to walk a mile in someone else's shoes and become more aware of different points of view. Be objective; most importantly, see yourself as others see you. Those with strong and difficult personalities will find it hard to relate until they correct their offensive behavior and irritating personality traits. If you are determined to get your own way, life will not run smoothly. You must consider the way you affect others. People will no longer make allowances for your behavior and it's time to clean up your act. If you can't learn to cooperate and share equally, you will have little chance for success and a greater tendency to create enemies. Correct those habits which block intimacy and sharing with others if you want to increase interpersonal gratification.

THE SUN IN THE 8TH HOUSE

This is often a year of tremendous change. It is common for individuals with an 8th house Sun to change their lifestyle completely during this year. There is a death of sorts associated with this house, but it is the kind of death that comes from strong transformation and change. Common examples are quitting work and returning to school full-time, leaving home and getting an apartment, transferring to another part of the country and living in a totally different climate and

neighborhood (e.g., moving from the city to a rural area). The emphasis is on radical change. Usually there is at least one major change during the year accompanied by many minor changes. Mental stress can result from the number of changes handled within a short period of time.

An increased perception on both the psychological and intuitive levels is common. During the year, you become more aware of subtle energies. Usually, individuals with this placement begin to feel that they see and know too much. They become overly perceptive of others' true feelings, motives and psychological inadequacies. For those who can use this new ability positively, this is a time of tremendous growth and enlightenment. Concurrent with this insight into human behavior is generally an interest in psychological and esoteric topics. Knowledge of this nature will help you to both understand and cope with your perceptions. Your own actions can become immune to manipulative ploys.

For those who do not understand or cannot appreciate their newfound abilities, this is a time of disappointment and helplessness. Unconscious reactions will take the place of rational interchanges as you respond from the gut level without thinking clearly. It is easy for others to push your psychological buttons and manipulate your behavior since stressful situations or relationships have a greater effect on you. This is more likely to occur if the air element is also lacking in the solar return chart. Examples of compulsive behavior or obsessive/phobic thought patterns will occur regularly. The less aware you are of your own drives, the more easily you will be controlled by them. The awareness of psychological complexes in others should make you more aware of your own psychological idiosyncrasies. Merely reacting to external forces is not as desirable as planning and directing your own life.

The whole purpose of this year is to gain power over yourself and your life. You may do this by understanding and overruling your unhealthy psychological impulses. If you don't, others can gain power by using your own unconscious against you. Guilt and jealousy are just two of the more common manipulative emotional tools others can and will use. You also gain power by recognizing when others are not thinking clearly. It is not necessary to respond to irrational demands. The first step to controlling psychological complexes is recognizing how they influence behavior. The second step is to become aware of why these counterproductive feelings and situations exist. The third step is to simplify your life by eliminating those psychologically unhealthy influences which block rational thinking and meaningful relating.

Obviously this is a good time for therapy, for you and/or for those you are involved with. People who seek therapy while the solar return Sun is in the 8th house generally only need counseling for a specific problem rather than deep psychoanalysis. Although some individuals have found this placement very difficult mentally, the greater number of mental illness cases are associated with the Sun in the 3rd and 12th houses, not the 8th.

Power struggles over money and morals are common, especially if there are also planets in the 2nd, 5th or 11th houses. Your conflict may involve another person or you may be struggling to control your own behavior. Spending practices usually change and debts either go up or down dramatically; they rarely stay the same. Many individuals rip up their credit cards and pay off their bills, but others take out a large loan. Generally, there is nothing in between these two manifestations. If you normally have high credit bills, you will feel the crunch this year and begin to cut back. If you are normally very cautious, you may be asked to take a financial risk because of the changes you are making.

THE SUN IN THE 9TH HOUSE

The Sun in the 9th house of the solar return chart indicates the need to reassess beliefs. The 9th house is not limited to higher thoughts only, but includes all beliefs, mundane as well as philosophical, nonreligious as well as religious. This is also the house of prejudice, intolerance and fanaticism. Beliefs that are erroneous, impractical or unsuitable need to be confronted and eliminated during the year.

Now is the time to reevaluate all beliefs, including those about yourself, your abilities, and relationships. If you have a long-standing belief in your ability to accomplish something difficult, this is the year to make the attempt. As you test your beliefs, one or more of them may prove to be a "misbelief" or misconception. Not all will be accurate or valuable. True beliefs must be proven so they can be built upon. They are the cornerstones of future actions. Misconceptions, on the other hand, can have serious consequences. They distort your perception of reality and stunt your emotional and spiritual growth. They must be corrected so that evolution will resume.

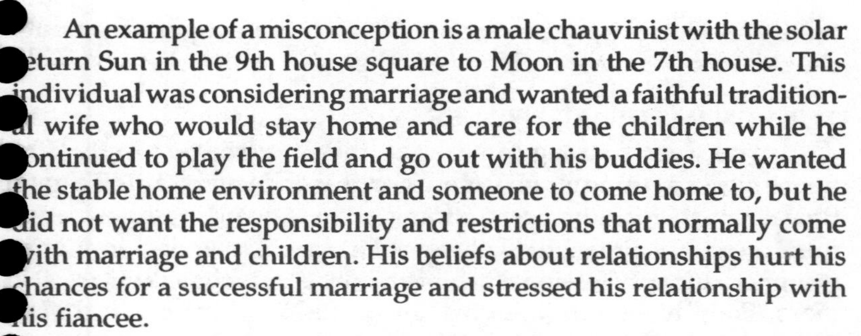

An example of a misconception is a male chauvinist with the solar return Sun in the 9th house square to Moon in the 7th house. This individual was considering marriage and wanted a faithful traditional wife who would stay home and care for the children while he continued to play the field and go out with his buddies. He wanted the stable home environment and someone to come home to, but he did not want the responsibility and restrictions that normally come with marriage and children. His beliefs about relationships hurt his chances for a successful marriage and stressed his relationship with his fiancee.

The spiritual philosophy which is the guiding force behind your everyday actions should be grounded in practical application. You should be able to live comfortably with your beliefs without the pain of continuing disillusionment. Some spiritual philosophies are too idealistic and self-defeating, encouraging the individual to remain in situations which are psychologically unhealthy. Turning the other cheek when you are being abused makes no sense. Your spiritual and religious beliefs should lead you toward fulfillment and peace. Philosophies which do nothing to improve the Universe and lack realistic manifestations may be meaningless mental exercises. A focus on coping with continuing disappointment shortchanges the individual. Use this time to find a philosophy that improves your life and the lives of others.

This is also the time to practice what you preach. If you are not aware of the inconsistencies in your beliefs, the contradictions will manifest themselves in your behavior. Others will notice the discrepancy between what you say and what you do. If you are truly hypocritical, you will tend to attract situations which accentuate this conflict even more.

This is a wonderful time for learning. If you have been out of school and wish to continue your education, do so now. Professional refresher courses will help you next year when the Sun moves into the 6th house of work. You can study any topic during the year and courses do not have to be part of a formal education. Concentrated study is likely and you may do this on your own or within a structured environment. Those who have completed their education may prefer to lecture or teach. As an alternative, you might sell a book proposal with writing and editing taking most of the year.

The Sun in the 9th can also indicate extensive traveling. You are more likely to travel outside the country and overseas, especially if there are other 9th house placements. Experiencing other cultures should make you more tolerant of others and their belief systems.

THE SUN IN THE 10TH HOUSE

If you want a strong career year, this is it. The Sun in the 10th house places a strong emphasis on all career endeavors. This is a good time to push for advancement and work towards greater career fulfillment. You are more likely to make major career decisions or be recognized for professional achievements at this time. You are also more likely to be reprimanded for professional misconduct. The reputation you have developed in the past catches up with you and it can be good or bad depending on what you have or have not accomplished and your methods along the way.

If you have positive feelings about your job and have worked hard all along, push for advancement. If you have negative feelings about your job, do not let them threaten professional goals. You are better off changing jobs or starting your own business than staying in a job you hate. If you choose to remain in a difficult situation, you will find work too taxing, physically, mentally, or spiritually. Personal needs will begin to conflict with the job requirements and you will feel pulled between home, work, and relationships.

If there are aspects into the 1st house, your personality quirks might not lend themselves to career tasks, thereby threatening advancement. You can handle these conflicting feelings positively by integrating your career with your personal needs and balancing all the areas of your life.

With the Sun in the 10th house, you have a lot of control over the direction of your professional life and also the direction of your life in general. Since the 10th is the house of destiny, you can make decisions about where you are headed for the next set of years. Pick a new path or life direction. Normally the focus of attention shifts to a new area of major concern, and this focus of attention persists for several years. Changes usually relate to or affect your career, but not necessarily.

Authority issues are strong at this time. If you are the authority figure at work, you should be very mindful of managerial techniques and concepts. Your professional success is directly tied in with your ability to lead and motivate those under your control. If you are not the person in charge, then you must learn to please those who are. They are the ones who recognize and reward your hard work.

If you are self-employed, success will come from your ability to perform well and please the public. In any of these situations, how well you cope with career tasks will determine the measure of success you achieve by year's end. Your interest in learning and mastering effective business techniques at this time will help you improve your chances for success.

Your parents can influence your decisions either positively or negatively, especially if you still see them as authority figures. If you consider either your father and/or your mother an authority in your field of interest, go to them for advice. You may make decisions in spite of them, especially if you feel the need to break any authoritarian hold they might have over you. This is the time to determine your own future. You are the final authority. Great opportunities wait for those who are prepared to meet the challenge and take responsibility for their decisions.

THE SUN IN THE 11TH HOUSE

Capricorn and the 10th house are associated with the laws and rules that are written for the masses. These laws structure society as a whole. Aquarius and the 11th house rule the personal reassessment of those laws, their meaning and application on an individualized level. Conscientious objection and a disregard for societal norms are also Aquarian themes. The 11th house in the solar return chart is the house of "Why not?" Why not have an affair? Why not start your own business? Why not sail around the world alone? Why not, why not, why not! The sky is the limit.

When your Sun is in this house, you must personally accept or reject all rules that affect your life. Rules are questioned, and either broken or found to have great personal validity. Morality and ethics are subject to personal scrutiny. You must decide what is right for you, your life situation and the people around you. This testing of rules and laws can be a mental exercise that lasts all year or it can be a difficult struggle involving strong urges. Those individuals with strong Uranian themes in their charts are more likely to act out. Those with strong Saturnian themes in their charts are less likely to act out but more likely to be upset by the review process. This is because strange thoughts can be associated with the Sun in the 11th house. These are not conservative, practical thoughts but wild crazy urges sometimes lacking a basis in reality. If you are not aware of the purpose for this process, you may be frightened by your own erratic thinking.

The important task for this year is to weaken Saturnian restrictions that are outdated and inhibit your growth potential. Next year your Sun moves to the 8th house where many changes will occur. The Sun in the 11th house indicates the need to prepare for these changes by reviewing meaningless patterns of behavior. By the end of the year, you should develop a personal code of behavior that makes sense within the context of your situation and allows you to make necessary changes.

The 11th house is also the house of future goals and hopes. This is the year to activate goals, or at least determine if they are realistic, practical or suitable as a future endeavor. Some dreams will survive the test, while others may not. Begin to work on those goals which are feasible and worthy of your attention.

Freedom is a major issue, and the amount of freedom you experience in your life will either rise or fall. If you have been feeling suppressed and limited, this will be your year to break those restrictions. You may actually forsake your former code of conduct or pattern of behavior. For this reason, the married individual is affair prone, especially if there are also placements in the 5th house. Friends may become lovers and lovers are friends and it becomes difficult to draw lines between the two. This is as true for single people as it is for those who are married. It is also possible that your freedom is seriously restricted as you strive to meet your goals or fulfill your dreams. Children and babies can be the fulfillment of a dream but at the same time, they do limit your freedom.

When the Sun is in the 11th house (the 2nd of the 10th), it is probable that your income will increase. Last year, if your Sun was in the 2nd house, you decided how much you were worth in the labor market. This year you set out to earn that much or at least increase your income. There is the possibility of advancement or promotion.

Group interaction will be important and this is a good time to experience or observe group dynamics. You are more apt to perceive subtle shifts in power within the group, or more obvious power struggles between members. You may or may not actively participate in these struggles yourself, but you will be aware of the power individual members possess and also the power the group possesses as a unit. Your experience can be mostly passive (in which case you will tend to go along with the crowd), or more aggressive (you strive for a position of influence). If you are interested in a cause or project, you should be able to harness group energy to accomplish the task at hand if your intentions are honorable. This is a good time to experience your own power and ego as expressed through a group. But if you waste this time on mindless in-fighting, all victories will be hollow.

And finally, the Sun in the 11th house relates to developing and also fading friendships. Those friends who did not treat you with respect last year are no longer around. New friends that you attract should be more appreciative. Relationships are less intense than the one-on-one partnerships of the 7th or 5th house, but no less meaningful. You can have in-depth but intermittent encounters with several friends or many superficial acquaintances. This year you are learning to express yourself to a variety of people and consequently your circle of friends expands. Both new and old friendships might be instrumental in the achievement of your goals.

THE SUN IN THE 12TH HOUSE

The Sun in the 12th house indicates that this is a behind-the-scenes year for you unless the Sun is close to the Ascendant. This is not a year in which you will seek to be noticed. Recognition is more likely to come next year and you can be noticed then for something you accomplish now. Socially, you will tend to be quiet, more withdrawn and preoccupied with internal thoughts or projects.

It is time for the important task of organizing information you have collected while the Sun was in the other cadent houses. Consequently, you will need time alone for reflection and introspection. You may daydream and fantasize a lot, or spend time contemplating your existence and the meaning of life. A religious theme is common, but all information is reorganized and not just data with a philosophical or religious theme. You will tend to keep your thoughts and feelings to yourself since they are usually only partially formed or partially understood for most of the year. You tend to be tolerant of different beliefs and will not push your incomplete opinions onto others.

Year-long goals and projects are often a direct result of the reorganization process. This is usually not the year for short-term success. You should be preparing for the future at least one year ahead. This is not to say that this is a wasted year; this is a excellent Sun placement for those who are working on a long-term project and do not expect to be rewarded or praised until they complete the task at hand. Generally, the fruits of their labor will not be evident until the start of the next solar return year. If the Sun should go to the 10th house the following year, this would be the time of career recognition and praise. If the Sun should fall into the 9th, this could be a time for publication. Authors who write their books with the Sun in the 12th house often present them to publishers as the Sun moves into the 9th.

If you are working, you may find it difficult to gain recognition now for the work that you are doing. Again, you might have to wait until next year when you complete a long-term project. But some individuals with this Sun placement work in a back room forgotten by management. They rarely deal with the public on a regular basis unless they are a voice over the telephone. The work they produce is unsigned and anonymous. A good example of this kind of work situation would be the publication of a newsletter that does not carry your name as writer or editor. You may hope to finally receive the recognition you deserve in the following year, but if you have your doubts, you might consider switching jobs.

If you are a housewife with children, you may prefer to spend more time at home. A demanding domestic situation will cause you to retreat from extra commitments and a hectic social schedule. You might have a major project you are working to complete, such as redecorating your home, but more simply, the demands of your family and children can be overwhelming. Numerous responsibilities will drain your energy and time alone helps you to recuperate.

This is also the year of the "closet personality." People will wonder where you have been and what you are up to since you will tend not to be self-disclosing. Perhaps your reputation is clouded. You might feel egoless. You are less likely to make demands on others and more likely to accept things passively, even though it is unusual for you to act this way. You can ignore feelings, thoughts or even desires for the sake of another person, especially if that person is in need. Negatively, you may be trying to bite your tongue and keep peace rather than defend yourself or say what is on your mind. If you are consciously trying to hide something, you can be secretive and deceiving, but it is also possible that others actively deceive you. It is more usual to be confused than deceptive or deceived; more positive to be enlightened than disillusioned.

This might be a year when you are involved with the sick or disabled. The individual with Sun in the 12th is usually a good

Samaritan. He or she spends time helping those who are incapacitated or confined, visiting friends, neighbors or relatives in the hospital. Compassion increases this year, but it is better to help those who are truly in need than those who refuse to respond or take responsibility for themselves. Beware of the savior-victim syndrome. This is a very negative manifestation resulting from intense involvement with alcoholics, drug addicts, or mentally unstable people. You may think you can help them, and they may look to you for assistance, but as the year progresses you see that they are slipping back into their old patterns and you feel drained, used and disillusioned. Steer clear of these kinds of people in personal relationships and use your compassion to direct them towards professional therapists. Working with these kinds of individuals professionally, however, may be very fulfilling for you.

This is a year that can be unstructured and Neptunian. Your goals and philosophy may evolve slowly as the year goes on; consequently, it will not be to your advantage to structure and plan your schedule far in advance. Some individuals will feel that they have little control over situations because the needs of others overwhelm their own needs and they are easily sidetracked. Some will not have good control over their minds either. Those individuals who were counseling candidates before the year begins usually find this to be a very difficult year. Mental afflictions and neurosis are possible for those who do not seek counseling when they need it.

Healthy individuals may notice more anxiety and nervousness as unconscious patterns surface and create difficulties. You might be phobic, worried, indecisive, less confident, or less assertive. If you are already in therapy and have been for a while, this is the time when you begin to see previously hidden patterns of behavior. You are able to put everything together and long-term issues finally become resolved.

It is important that you get proper rest and nourishment. When the Sun is in the 12th house, you may be more easily drained physically and emotionally. You need time alone for contemplation and planning; illness is one way to get it. This is more likely to happen if you are overwhelmed by numerous responsibilities which you do not allow yourself to avoid. Learn to say no. Time spent alone in reflection can give you a renewed sense of faith in the Universe and a new sense of dedication to the role you play in its creation and evolution.

CHAPTER THREE

THE MOON IN THE SOLAR RETURN CHART

Introduction

The solar return Moon seems to have two consistent themes throughout all of the houses. First and most simply, the Moon is associated with change and fluctuation according to its house position. Secondly, and probably more importantly, the Moon reflects the individual's emotional nature during the solar return year. Feelings, temperament and emotional needs can be assessed by the Moon's placement, sign, and aspects. Certain themes normally associated with the Moon are not consistently related to all of the houses. For example, themes concerned with the home may be seen with the Moon in the 2nd or the 4th house, but generally not when the Moon is in the 7th.

The Moon and the Cycle of Change

Change as a natural process is normally implied by the Moon. We think of ocean tides and lunar cycles as natural changes occurring in a rhythmic pattern. The Moon in the solar return also seems to indicate a natural cycle of change.

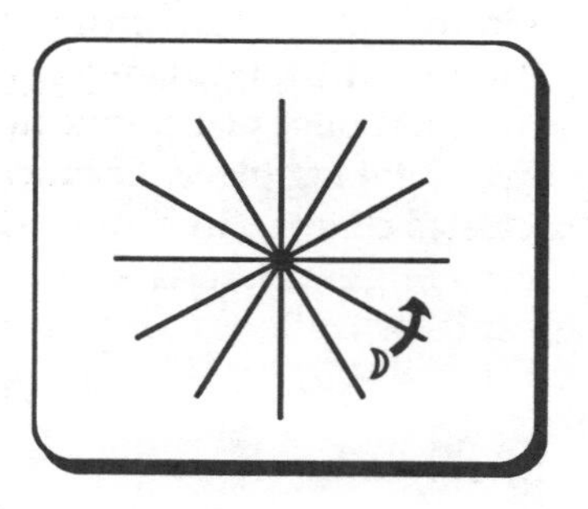

Figure 17

The Moon moves through the solar return charts in a steady counterclockwise progression, moving approximately one or two houses per solar return year. If your Moon is in your solar return 5th house this year, it was probably in the 3rd or 4th house last year and may be in the 6th or 7th house next year. (See Figure 17.)

Even though the Moon changes degrees each year, approximately 100 degrees forward in the zodiac, it is sometimes possible for the solar return Moon to remain in the same house for two years in a row. The Moon skips some houses entirely. The Moon's cycle around the solar return chart takes approximately eight years. This rhythmic pattern occurs if the individual remains in the same location from year to year.

This lunar cycle suggests that change occurs in a set sequence, and the sequence is understandable. For example, if you have the Moon in the 1st house this year, it will probably be in the 2nd house next year. This pattern logically implies that you must become aware of your own specific emotional needs (Moon in the 1st house) before you stand up for yourself and make demands. Self-worth issues (Moon in the 2nd house) are supported by an accurate assessment of your needs and whether or not they are being met by others. Suppose your Moon now moves from the 2nd house to the 4th house. The need for financial security recognized while the Moon was in the 2nd may lead you to purchase a home as an investment (Moon in the 4th house). If instead the Moon moved from the 2nd to the 3rd, a lack of security in your old home may prompt you to investigate new, safer neighborhoods (Moon in the 3rd house).

Each year is a continuation of the last year, and last year's successes become this year's foundation for future advancement. In the same way that the zodiac proceeds from Aries baby to Pisces

Good Samaritan, the Moon in the solar return chart proceeds from emotional self-knowledge (Moon in the 1st) to Universal compassion (Moon in the 12th). The Moon by its house placement indicates where evolutionary emotional growth is taking place. The individual experiences this maturation process as changes in circumstances. Unstable situations and fluctuating feelings are likely, but normally changes are minor. Major changes and disruptions are more often suggested by Uranus and/or Pluto.

The Sun-Moon Cycle

The Sun and Moon have a constant relationship cycle known as the eclipse cycle. Eclipse degrees rotate through the zodiacal signs, completing one cycle every nineteen years. The pattern is then repeated. Therefore the lunar placements in the solar return also cycle for nineteen years, creating a series of nineteen points. If you were to make up solar returns for a block of nineteen years in a row, you would have the approximate positions for all of your solar return Moons for your life span. After the Moon has cycled through the initial list of placements once, it will begin the cycle again, falling on the same points, plus or minus a few degrees.

Movement of the Sun and Moon

The Sun and the Moon both symbolize separate and distinct patterns of growth within the solar return. Together they seem to form a more complex pattern for maturation. The Moon's movement in the solar return is in sharp contrast to the Sun's since they circle the chart in opposite directions.

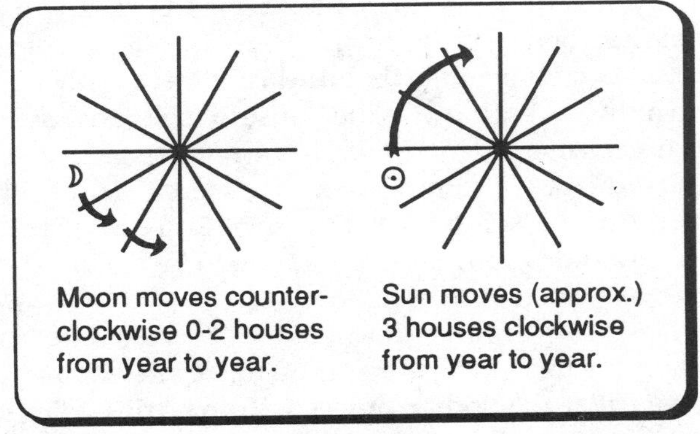

Figure 18

The Sun and the Moon in the solar return are like two ships passing in the night. If you remain in the same location from year to year, the Sun will move around the solar return chart in a clockwise progression moving approximately 3 houses each time. (Some distortion may occur with signs of long or short ascension.) The movement of the Sun is very closely tied to the changes in the Midheaven, as explained in Chapter 2 of this book. It takes four years for the Sun to circle the chart. On the other hand, the Moon moves in the counterclockwise manner explained above. So you see, the Sun and the Moon are traveling through the solar return charts in opposite directions, each working within its own cycle.

The Moon's movement is particularly significant when you realize nearly all of the planets move clockwise around the chart. Those that are associated with the Sun, such as Mercury and Venus, must keep pace with the Sun and therefore move clockwise. The outer planets, from Jupiter to Pluto, are too slow to beat the rapid rotation of the solar return chart; consequently, they too will appear to be moving in a clockwise direction. Mars is unique in that it appears to shift erratically around the chart with no consistent pattern whatsoever. The Moon is the only planet which bucks the tide and travels opposite to the Sun. The two lights, each traveling through the solar return in opposite directions, symbolize very different patterns of growth. Because of their converse movement, it is especially important when they eventually meet in the same house; the combination can signal a very significant year.

There are three possible combinations for the Sun and the Moon placed in the same house. First, they may have no relationship other than being in the same house; second, they may be conjunct within a ten degree orb; or third, the Moon may eclipse the Sun. Generally the closer these two planets are in the same house, the more intense the union. Solar returns on a solar eclipse point are especially significant. (See figure 19 on page 74.)

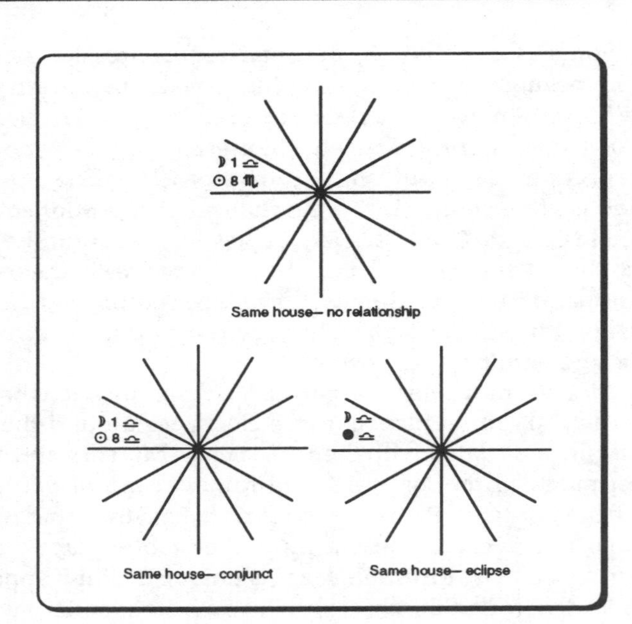

Figure 19

The Sun and the Moon in the Same Solar Return House

Having the Sun and the Moon in the same house is like having all of your eggs in one basket. The Sun is associated with conscious experience and symbolizes a need for external achievement and recognition. The Moon, on the other hand, is more closely associated with unconscious experience and the need for internal emotional fulfillment. Whenever they appear in the same house, there is a conscious and unconscious emphasis on the same area of life. Matters related to the house placement are very prominent during the solar return year.

The convergence of the two lights in the same house symbolizes the merger of two separate and distinct patterns of growth and the formation of a new, more complex pattern based on the meeting of the Sun and the Moon. It seems reasonable to speculate about the existence of this complex cycle since whenever the Sun and Moon appear in the same solar return house, there is a sense of "newness" attributed to the events of the year. Significant (Sun) changes (Moon) can

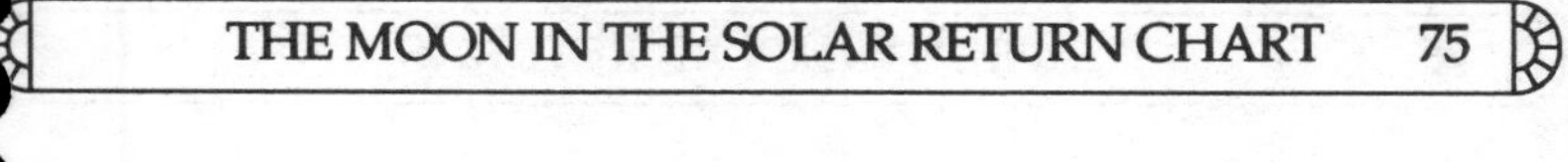

occur, and these changes may be so far-reaching as to signal the beginning of a new era. It is during this year that the individual is involved in a new experience and/or new issue. Generally, the closer the Sun and the Moon are within the house, the greater this sense of newness and the greater the sense of agreement between the conscious and unconscious levels. The farther apart the Sun and Moon are placed, the more likely one is to experience two separate desires or needs.

For example, if the lights are not conjunct and not even in the same zodiacal sign, you are of two minds concerning the subject matter of the house placement.

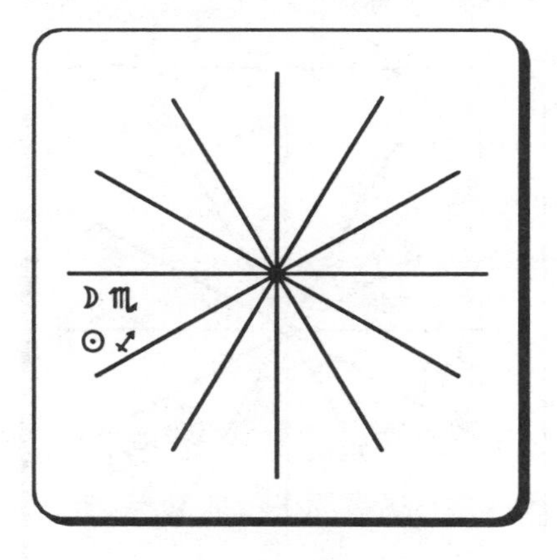

Figure 20

A new divorcee with this configuration was on her own after many years in a restrictive and unsatisfying marriage. She loved her newfound freedom (Sagittarius Sun in the return 1st house), but she also loved the intimate and sexually fulfilling relationship she now had with her new boyfriend (Scorpio Moon also in the 1st house). During the solar return year, the question of marriage came up many times. The woman did not want to chose between two equally desirable situations. She wanted to fulfill both her need for freedom and her need for intimacy. The fact that the Sun and the Moon are in different signs suggests that your conscious thoughts on the matter at hand will be different from your unconscious feelings. If this is true, you are apt to feel pulled in two directions. However, if the Sun and the Moon are in the same sign, you are more likely to experience conscious-unconscious agreement.

Sun-Moon Conjunction in a Solar Return House

When the solar return Sun and the Moon are conjunct and in the same sign in the same house, you are more likely to experience consistency. And as the lights come closer together, the sense of "newness" becomes especially noticeable. Generally, the new experience is related to the house placement of the Sun and Moon, but if the native is an adult, the new experiences may occur in the career or professional arena regardless of the Sun-Moon house position.

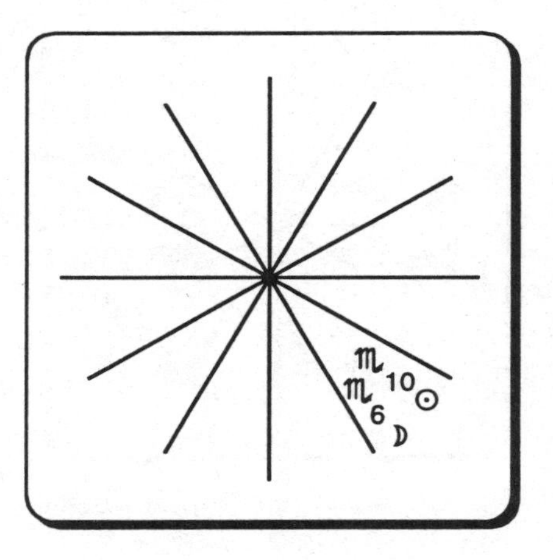

Figure 21

For example, a teenaged girl with the Sun and the Moon in the 5th house had her first real boyfriend. He was much older than she and brought up the issue of having sex. Both the relationship and the sexual issue were new experiences. She went from girl to woman very quickly. Each of the lights was in the 5th house, and the two were conjunct.

Solar Eclipse in the Solar Return

A solar eclipse occurring on the same day and near the time of the Sun's return is a fairly uncommon event. Check your nineteen solar

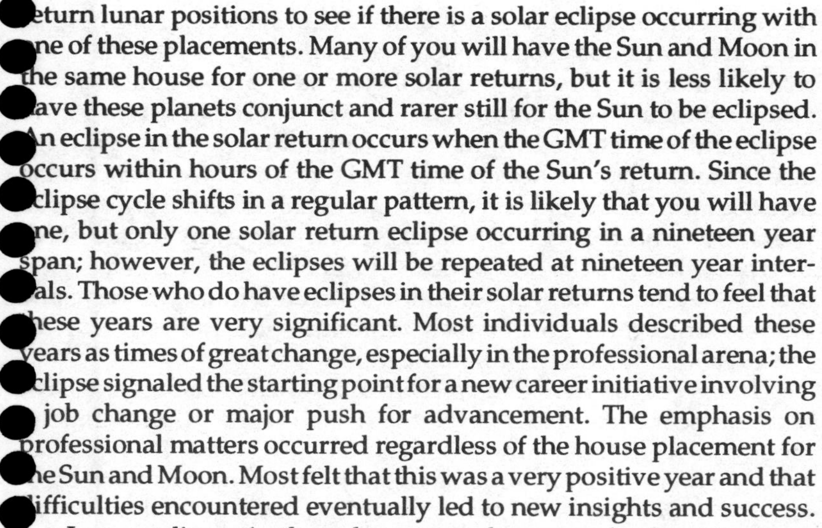

return lunar positions to see if there is a solar eclipse occurring with one of these placements. Many of you will have the Sun and Moon in the same house for one or more solar returns, but it is less likely to have these planets conjunct and rarer still for the Sun to be eclipsed. An eclipse in the solar return occurs when the GMT time of the eclipse occurs within hours of the GMT time of the Sun's return. Since the eclipse cycle shifts in a regular pattern, it is likely that you will have one, but only one solar return eclipse occurring in a nineteen year span; however, the eclipses will be repeated at nineteen year intervals. Those who do have eclipses in their solar returns tend to feel that these years are very significant. Most individuals described these years as times of great change, especially in the professional arena; the eclipse signaled the starting point for a new career initiative involving a job change or major push for advancement. The emphasis on professional matters occurred regardless of the house placement for the Sun and Moon. Most felt that this was a very positive year and that difficulties encountered eventually led to new insights and success.

Lunar eclipses in the solar return chart can also be important. Career changes are sometimes associated with the lunar eclipse as well as the solar eclipse. The interpretations can be more varied, though, and may reflect the house positions for the Moon as well as the Sun. On the whole, lunar eclipse solar returns do not appear to be quite as significant as those with solar eclipses.

The Moon as an Emotional Indicator

While we speak of the Moon as simply an indication of one's emotional nature, the actual interpretation of the Moon is more complex than simple. The Moon is not just an indicator of feelings; one's emotional nature is a multidimensional conglomeration of feelings, temperament, needs and unconscious responses. We cannot look at the Moon as one-sided; we must study its interpretation from a number of perspectives.

The Moon represents the individual's feelings regardless of whether or not they are expressed openly. The type of feelings you have during the year can be demonstrated by the Moon's sign, position and aspects. The Moon's sign can be important even though the Moon has essentially only nineteen placements, because it is indicative of what the individual feels during the year and how he or she is most likely to exhibit those feelings. For example, a solar return

Moon in Cancer, interpreted by itself and out of context, implies that the individual tends to be more emotional during the year. He or she may feel more nurturing towards others, and therefore tend to attract relationships and situations which allow the expression of protective or dependent feelings. On the other hand, the individual with the solar return Moon in Aquarius could be preoccupied with mental activities and not be concerned with emotional issues during the year. Certainly, placements and aspects modify or even negate these interpretations, but by and large they will be true. The Moon sign is a valuable clue to the types of feelings the individual will have and express during the solar return year. It is also indicative of the individual's emotional temperament.

The major difference between an individual's feelings and basic temperament is that feelings tend to make up one's basic temperament. The word temperament implies one's overall pattern of emotional response; it is an emotional factor that remains constant regardless of the circumstances. On the other hand, feelings imply a specific emotion in response to a particular event or situation. For example, moody people have emotional swings. Their feelings change depending whether or not they are experiencing happy or sad events; however, their proclivity towards emotional highs and lows remains constant regardless of their particular mood at any point in time. Despite feelings of joy or depression, they still have a moody disposition. The solar return Moon's sign is generally very descriptive of your emotional temperament. It can suggest an array of emotional characteristics including moodiness, sensitivity, coldness, or enthusiasm. As always, the interpretation is modified by the Moon's house placement and aspects.

Emotional needs play a crucial role in the Moon's cycle of change and emotional maturation. As explained above, the Moon moves in an understandable pattern through the solar return. During each year, the individual knows certain emotional needs must be met in order to feel secure and fulfilled. These needs, which fluctuate from year to year, cause the individual to create the variety of situations necessary for emotional maturity. The Moon's sign, aspects and placement suggest what type of experiences will be needed in order to continue the maturation process.

Conscious or Unconscious Emphasis

The Moon also signifies the unconscious experience. While Mercury's placement will suggest what the person is consciously thinking, the Moon's placement will suggest what the individual is experiencing on an unconscious level. Together, the Moon and Mercury (and sometimes also the Sun) symbolize the balance between conscious and the unconscious processes. We like to think that our decisions are based on the integration of rational and emotional information, but during any given solar return year, one process more than the other can influence our decisions, perceptions, and motivations. The focus of attention shifts to accommodate fluctuations in individual patterns of growth. There will be years when you lean more heavily on your emotional responses, tending to trust your instincts and follow your intuition. But there will be other years when you depend more on a logical assessment of your situation. The conscious-unconscious balance in the solar return chart can be understood by assessing the Moon and Mercury (their positions, signs and aspects) and various other rational and emotional indicators.

Certain astrological symbols are normally associated with an emphasis on the unconscious mind. Included here is a listing of the most likely solar return indicators for this emphasis. There are a number of different ways in which the unconscious process can be emphasized more than rational thought. Please keep in mind that it is the cumulative emphasis on the unconscious that is important and not the presence of any one factor. A solar return chart which shows an emphasis on the unconscious mind would tend to stress the Moon and/or Pluto (by sign, house or aspect), the water element, the water houses, and a preponderance in the northern hemisphere. Mercury placed in a water sign, or in a water house, and in major aspect to Neptune, Pluto or the Moon is another indication. The emphasis on the unconscious usually indicates a year in which the native will be interested in fostering emotional connections and unconscious insights. Logic and factual information may not be meaningful within the context of present situations, which will tend to be emotional. Example situations include (but are not limited to): involvement in counseling, therapy or self-help discussion groups; intense family situations (related to birth, child-rearing, sickness or death); and very nurturing social situations (e.g., communal living, helping the disabled, underprivileged or homeless). Emotional interactions and perceptions will naturally be more important if they are more likely to provide vital information needed to handle experiences during the coming year. An emphasis on the unconscious nature is conducive to emotional growth.

An emphasis on the conscious mind is usually seen with the following: a preponderance of the air element; all planets above the horizon and not in the 8th or 12th houses; the Moon and Mercury in an air sign; the Moon in the 11th house; Mercury in the 1st, 3rd, or 11th houses and not in aspect to Pluto or the Moon. Again, it is a combination of factors that is most important and not the presence of any one factor by itself. All forms of mental activity are associated with an emphasis on the conscious mind, including educational and informational activities. Teaching, learning, lecturing, and writing are several examples. An emphasis on the conscious mind is conducive to intellectual growth.

Although the Moon is probably the solar return's most important indicator for the individual's emotional nature, it is not the only one. Other planets are suggestive of emotional factors and can contribute to our understanding. Pluto reflects unconscious motivations, psychological complexes and intense feelings. Negative feelings such as resentment and never-ending anger can be suggested with Pluto. Neptune is associated with one's ability to be compassionate and sensitive to the needs of others. Mars implies our ability to be aggressive, assertive and angry. One feels motivated to take the initiative.

Mental Illness

Mental illness, anxiety, nervousness, and stress can be associated with all kinds of planetary combinations and are not limited to any one factor in the solar return chart. Minor problems, such as mild depression and/or nervousness, are common personality defenses. They are especially difficult to pick up in the solar return chart. However, once mental illness is already occurring, one can make a few correlations to the solar return for some of the more difficult mental problems. An extreme emphasis on either the conscious or the unconscious mind can suggest either suppression or overwhelming emotions respectively. If the person is mentally healthy, this shift in itself will not be important. Depression, especially severe depression leading to therapeutic intervention and suicidal tendencies, seems to

is most closely associated with the solar return 3rd house, and more specifically the Sun (or sometimes Saturn) in this house. Agoraphobia, a rare mental condition that makes individuals afraid to venture outside their homes, is more likely to be seen with an emphasis on the 4th or 12th house and the presence of the Sun in either of these houses. This is especially true for the year of the onset of the disease. Extreme nervousness and anxiety can be symbolized by the Sun, Uranus, or Neptune in the 3rd, or placements in the 1st, 4th and 12th houses. Compulsive-obsessive tendencies are commonly denoted by the Sun in the 8th house or the Sun or Moon in strong aspect to Pluto. Please keep in mind that if mental problems do occur, you may be able to see these placements in the solar return. The reverse is not true; these placements by themselves are not automatic signals that mental difficulties will necessarily occur. If you read the placements of these planets in the houses, you will see that there are positive options for all of these positions. They are signs of great growth for those who deal with life in an honest and open way.

Planets Aspecting the Moon

Aspects to the Moon can imply that emotional expression is either enhanced or inhibited by various personality factors and life situations. The planets aspecting the Moon, of course, color its interpretation.

MOON-PLUTO ASPECTS

The Moon in aspect to Pluto implies an emotional make-up which is complicated by unconscious psychological influences. Emotions may be tainted or seasoned by events or complexes from the past, and you could be reliving a former happening in an immediate situation. Feelings seem more volatile, intense or even overwhelming. Reasoning can be based on emotional factors and simply defy logic.

Relationships are likely to be affected and psychological influences will distort communications. This may be a time when you are in close contact with someone who is not always rational and you must deal with issues in an insightful way. In extreme cases, the other person needs counseling. This is a good time to become more aware of manipulations and psychological games. You may merely perceive these influences or you may be directly involved, playing either manipulator or manipulated.

Moon-Pluto contacts, at their worst, imply emotional power struggles for control. You may think you are struggling with another person, but really, you are only struggling with yourself. For example, suppose you are a woman of limited means married to a wealthy man who is very tight with his money. Since you resent your husband's power over your economic situation, you might decide to withhold sexual relations. You have now gained some power over your husband's life and the two of you have established a power struggle which complicates your relationship. Although you have succeeded in controlling your husband's behavior to some extent, you have not gained any power over your own existence. Your situation has not improved much and you are still without financial recourse. There are more positive ways to handle this situation. You and your husband could decide to seek marriage counseling to discover why you need this impediment to emotional intimacy. You could enter therapy yourself to analyze why you choose to remain with such a stingy person. Or you could get a job, earn your own money and totally frustrate your husband's attempts to control you.

Emotional blackmail and manipulation only work if you allow another person to have control over some portion of your life. As soon as you begin to control yourself and take responsibility for your own well-being, manipulators lose all power.

Not all Moon-Pluto contacts involve difficult circumstances. Writers, counselors, psychology students or those whose work depends on their ability to understand human nature tend to have Moon-Pluto contacts in their solar returns during important years. It is the awareness of the psychological and emotional influences which is significant and not the struggle itself. Knowledge brings power. Use this time to become aware of how unconscious drives affect your life. A very pleasant reason for being so emotional is an involvement in a new and exciting romantic relationship. All relationships, even those which are established, are subject to change and new insight. Domestic situations are also in the process of changing. Moving or planning to move is possible.

MOON-NEPTUNE ASPECTS

Moon in aspect to Neptune can show increased sensitivity to life's subtleties. While Pluto-Moon contacts show a greater insight into psychological influences, manipulative games, and a need for self-control, Neptune-Moon contacts indicate a greater sensitivity to feelings and needs that are not expressed openly yet bind us all. There are both positive and negative manifestations suggested by Neptune aspects to the Moon.

Negatively, you can be misinformed or even lied to, especially if someone is actively trying to deceive you. However, it is more likely that you will deceive yourself when there is someone important in your life whom you do not truly want to understand. This other person may be a family member or someone you are emotionally involved with. Neptune's house position will give you a clue. If someone's actions mystify you, making you feel the relationship is insecure, it's very likely that you do not have a realistic perception of this person, and possibly do not want to know the truth. The definitive truth might be more threatening than the mystery, so you allow the situation to remain clouded and do not challenge misconceptions. Since much is left unsaid and asssumed, you try to fill in the gaps with your expectations rather than realistic assessments. Disillusionment is always built on dreams rather than the truth. You may not see things as they really are, but only as you hope they would be. This continuing lack of accurate information leads to further confusion and fantasy, and is compounded by emotional estrangement. You, yourself, may not be in touch with what you really feel. Anxiety and excessive worry can be signals that your perceptions are not an accurate reflection of reality. It is perhaps fair to say that intuitive insights were never meant to augment expectations. This may represent a misuse of this newfound sensitivity.

Moon-Neptune aspects are meant to symbolize intuitive insights, spiritual ideals and one's greater sensitivity to the commonalities we share. We all suffer from human frailty; Neptune symbolizes our ability to identify with others, and see their failings as well as our own. True understanding supersedes any judgmental attitudes and paves the way for a meaningful exchange. Idealism and putting others on a pedestal diminish realistic achievements which might have been accomplished in spite of human weakness. Idealism and unrealistic expectations are defenses used to avoid facing frailties and humanness we do not wish to see.

Positively, Moon-Neptune aspects imply connections; connections to each other and to the Universe as a whole. We can foster and improve these connections by understanding, accepting and helping each other to progress as a whole towards a better existence. This is a good time to accept people as they really are, weaknesses and all.

You will probably be involved in situations which give you the opportunity to increase your understanding or insight into others. As long as you do not make demands or have unrealistic expectations, you should be able to retain honest emotional contacts with those around you.

You may actually care for someone who is ill or disabled, or you could simply care for another person more than you care for yourself and your own welfare. Moon-Neptune aspects are a sign of self-sacrifice. It is common to see this aspect in the charts of new mothers. The demands of caring for an infant involve a certain amount of self-sacrifice. If you work full-time to put your spouse through school, you might also have this aspect.

Self-sacrifice does not necessarily mean martyrdom. Moon-Neptune aspects in the solar return indicate that you are able to help others because you truly understand their situation. You are also able to handle emotional uncertainty. You accept relationships which are not clearly defined, and you are able to give without a guarantee of return. Going with the flow means accepting insecurity as a natural by-product of your situation. With Moon-Neptune aspects in your solar return chart, you need to be able to function during times of uncertainty and handle the insecurities that go with them. You need to develop trust.

MOON-URANUS ASPECTS

Probably the most common activity associated with a Moon-Uranus aspect in a solar return chart is moving from one home to another, especially when the Moon is conjunct, square or opposed to Uranus. You may move yourself, or help someone either move in or out of your home. Domestic changes are likely. Usually, a certain amount of disruption transpires in the home. Expect breaks in your daily routine and habits. Most likely, relationships are also changing. You could be involved in a new relationship, or an old one may be going through a period of transition or separation. Either you or your partner initiates these changes.

Strong attractions are possible at this time, but this is not the most common manifestation; romance is more closely associated with placements in the 5th house than Moon-Uranus aspects specifically. If you feel you are caught in an emotional rut, changes are more likely to occur. The external changes you experience in relationships are the direct result of internal restlessness. It stands to reason that if you situation and emotional needs change, you will be faced with new issues and problems, and be required to develop new ways of expressing yourself emotionally. This is one way to break inhibitions. The involvement in new situations will accentuate your ability or inability to handle emotions and relationships in a positive way. At the very worst, emotional control will be difficult and feelings will be erratic. You may be overwhelmed one day and detached or cool the next. You could say or do things without really considering the emotional consequences, but then, this might free you up to make necessary changes quickly and easily.

MOON-SATURN ASPECTS

While the Moon in aspect to Uranus suggests an inability to suppress emotional information, Moon-Saturn aspects are more closely associated with emotional control. If you are involved in an important project which you feel must be completed, you can work despite any emotional strain. You take your commitments seriously and will enforce restrictions on your own behavior or suppress feelings in order to get a job done. This is a time when responsibility wins out over emotional expression. For example, a high school senior pushed himself in his chosen sport. He practiced when he was tired and missed a number of social events during his senior year, but he was able to set a goal and work towards it. He knew the importance of his talent and worked hard to develop his potential fully.

Feeling that you can only depend on yourself and must take responsibility for your own welfare is characteristic of Moon-Saturn aspects. This is a time when you will be asked to make decisions affecting you emotionally. These decisions will have some bearing on your future sense of fulfillment (such as attending the school of your choice, or entering a particular career). You might have to make these decisions on your own if others are not involved, supportive, comforting or interested. They are either absent or for some reason unable to help you with the task at hand. At times, you may feel lonely. If you cannot depend on others, depend on yourself, your own resources and abilities.

Someone who started her own day care center had a Moon-Saturn aspect in her solar return. She worked alone to set up the center since she did not have an assistant. Her husband was fearful of the financial commitment and was less than supportive of her new business venture. There were times when she wished she could share her fears with someone who really understood.

Major decisions are made only after careful consideration, and may be associated with some sadness or stark realization. Emotional implications are likely. You may have to give up something to get something. Options may be limited, and you might have to choose from an either/or situation. Suppose you wish to move to your own apartment. Although you wish to move quickly, the particular apartment you wish to rent might not be available until later in the year. You can either wait or settle for something else. In the long run, the delay could be beneficial.

MOON-JUPITER ASPECTS

Moon in aspect to Jupiter suggests that your ability to express your feelings within a significant relationship is important to your own emotional growth during the year, and also to the growth of the relationship. Jupiter symbolizes your ability to expand and the Moon symbolizes your feeling nature. Together in aspect, they emphasize the process of sharing emotions, having those emotions understood, and growing from the interchange. For example, if you are involved in counseling, this exchange is crucial to the therapeutic process. If you are involved in a serious relationship, sharing your feelings is very important for the development of greater intimacy. Problems expressing your feelings or reaching a level of understanding will cause emotional stagnation and eventually strain your relationship.

A woman with a 12th house Moon in aspect to Jupiter was afraid to express her feelings to her lover because she wanted to get married and he didn't. She feared saying anything at all, sensing that he would leave her, and so kept quiet. Eventually, she also stopped expressing positive emotions such as love and affection. Whether she or the relationship was the cause of the blockage made no difference in the long run. Her inability to openly express what she was feeling caused her great stress and ultimately strained the relationship.

Some real or imagined barrier to expressing feelings can exist with this aspect, and it is important that you break through the barrier and be understood if emotional growth is to occur. Sometimes the barrier is really no problem at all, but a wonderful opportunity for openness. Expansion through emotional interchange and relatedness is the key. It can come about by overcoming barriers or creating rewarding opportunities.

Moon-Jupiter aspects sometimes imply overwhelming emotions. You do best with objective feedback from others. Your perception may be distorted if you keep your feelings isolated. It is just as important for you to convey cheerful emotions as serious feelings, especially if you are in a situation where enthusiasm could motivate others. For instance, if you are working on a humanitarian project and you need volunteers, your ability to inspire others with your enthusiasm will help your cause.

MOON-MARS ASPECTS

Moon-Mars aspects imply involvement in a situation which is emotionally uncomfortable yet rewarding, a mixed blessing. The energies represented here suggest working at cross-purposes and you may have ambivalent feelings about your circumstances. An example for illustrative purposes might be found in the experience of trying to comfort a sick baby. You know the baby is sick and does not feel well so you do everything you can to comfort the poor child. But at the same time, you become very frustrated if nothing you do seems to stop the screaming. At that point, stress is competing with concern and mixed feelings arise.

Mars in aspect to the Moon is very suggestive of contradictory emotions. The Moon symbolizes our nurturing qualities and emotional nature. Mars, on the other hand, symbolizes our independence and assertiveness. The basic interpretations for these two planets are so different that they tend to remain separate like oil and water. They represent two sides of the coin or different perspectives on one situation. For instance, a common situation often involving mixed emotions is having to say good-bye to a good friend because one of you is moving. If you are the person left behind, you may feel happy for your friend, but still sad about your loss. If you are the person moving, you may be excited about the move, yet sorry to be leaving. The purpose of the mixed feelings is to call attention to your own emotional needs (Moon) and take appropriate action accordingly (Mars). The mother of the fussy infant may need to take five minutes away from the baby to cope with stress. The person who is moving may need to concentrate on maintaining friendships even after relocation.

Moon-Mars aspects suggest that emotional situations are not simple and usually involve a strange combination of contradictory emotions. They can also indicate that you might act in a way that detracts from your own sense of security and well-being, thereby creating stress. An example could be involvement in an extra-marital affair. A Moon-Mars combination is not necessarily difficult, nor must it involve your personal life. It can refer to professional endeavors or social activism. You can be motivated to act because you are affected emotionally by what you see. If you are upset by conditions in a poor section of town, you might be motivated to volunteer your services at a local soup kitchen. The unpleasantness you feel impels the need to look at your surroundings more closely and work to correct or improve conditions.

MOON-VENUS ASPECTS

The Moon in aspect to Venus suggests that emotional security and domestic issues are tied to financial considerations. Your monetary situation can be affected either positively or negatively. If you are planning on pursuing an emotionally fulfilling goal, you will have to consider your financial situation before making any changes. Whether or not your goal is feasible at this time depends on how much money you have set aside. You might have to compromise your present income and take a pay cut to enter a more rewarding field. Ultimately, your salary will increase if you are successful in your new endeavors. Financial stability is also affected by domestic conditions. Moving, pregnancy and birth are common events that affect the amount of money earned and spent. Occasionally this aspect can show a time when you are financially dependent on others for support for one of the reasons mentioned above.

MOON-MERCURY ASPECTS

Moon-Mercury aspects emphasize the need for integration of unconscious feelings with conscious thoughts. If these two avenues for information and analysis are working together, they form a great combination; the integrated psyche is a powerful tool for intellectual and creative endeavors. You will be able to understand the total picture from your rational assessment combined with intuitive insight. When the conscious and unconscious are working together, you are more likely to make good decisions which satisfy your emotional needs. The conscious mind can be used to channel unconscious feelings into creative projects. Channelling can also be very therapeutic.

You can actively seek out more information about your feelings and motivations through discussions. But when the conscious and the unconscious function separately, the native experiences two distinct and contradictory pieces of information. It is often difficult to make decisions since you arrive at two mutually exclusive conclusions. You could consciously override or suppress your feelings and emotional needs. On the other hand, you may react without thinking. The key to using this aspect positively is to balance and integrate conscious and unconscious input and work towards a complete sense of self. (For more information see "Conscious or Unconscious Emphasis" above.)

The Moon in the Solar Return Houses

The Moon's house placement suggests the area of greatest emotional concern. The individual is most sensitive to the situations associated with the Moon's position. While the Moon sign can show how sensitive the individual is to emotional issues, and the aspects tend to indicate how effective the individual is in expressing feelings, the Moon's house indicates what the individual is most concerned about. The Moon shows the kind of emotional experiences the person is most likely to attract and need during the solar return year.

THE MOON IN THE 1ST HOUSE

The Moon in the 1st house can indicate increased emotional self-awareness during the coming solar return year. Feelings and emotional needs will seem stronger and more urgent, residing just under the surface. Sometimes the emotional nature is overwhelming and cannot easily be controlled. Individuals with this placement tend to cry easily, and to feel deeply. They may also be socially withdrawn or quiet. There are several other lunar placements which are associated with overwhelming feelings (i.e., the Moon in the 4th or 8th house); but the Moon in the 1st house is distinguished by the fact that not only are the feelings particularly strong, they are also readily apparent to others and ultimately carry great personal significance for the individual. Resolutions for the future are made from the subjective interpretation of these pressing feelings. Feelings do not just gain strength during this time, they become a motivating force. Feelings become useful. They have rhyme and reason within the context of events and they play a major role in decisions that you make. This is not the time to make totally rational decisions without considering the emotional ramifications.

For example, a new mother with the Moon conjunct Saturn in the 1st house of the solar return chart promised to return to work after the baby was born. Her full-time income was needed to support her student husband. She did not anticipate any problems with this arrangement, but she had not assessed how she would feel about the set-up. She had made an intellectual decision without any emotional input. Once she returned to work, she realized how unhappy she felt leaving her baby every morning. As the year wore on, she and her husband made decisions about their future that changed their work schedules significantly. The following year, the mother was able to stay home full-time to care for their baby. Her feelings about working while her child was young strongly influenced the choices she and her husband made. After this woman interpreted her feelings and grew to respect those emotions, she was able to work towards a positive resolution of the issue.

When the Moon is in the 1st house, your emotional nature is trying to tell you something important. Your feelings can be viewed as a connecting link between events in the past and guidelines or preparation for the future. It is common to find the Moon in the 1st house of the solar return when one is recovering from an emotional trauma, regardless of when that trauma occurred. It could have occurred in the very recent past or it could have occurred long ago (especially if the individual has a habit of suppressing feelings). Or you may be involved in a situation or personality pattern that is detrimental to your emotional nature or development. Usually this situation or pattern is well within your power to change. Present events trigger feelings, both old and new, and force them to the surface. This is not meant to be a depressing time, though sometimes this is the case, especially in the early months of the solar return. Moodiness is a common complaint. The goal is to get in touch with your feelings and listen to them. Understand what they are saying to you. Some pains are inevitable and some pains are avoidable. You cannot change death, but you can change the way you handle emotions, situations and relationships. The Moon in the 1st house signals a very strong developmental stage for the feeling nature. This is not just a year of recovery, it is a year for great emotional growth.

This is also a time to become self-protective and self-nurturing. The Moon in the 1st house could be titled, "The Care and Feeding of your Emotional Nature." Many of us do not know how to soothe our own injured feelings and heal old wounds. Use this time to learn how to protect yourself from emotionally damaging situations while recognizing and fostering emotionally healthy relationships. Appreciate your own emotional needs and know when they are not being met. Allow your feelings to play a greater role in the decision-making process.

This is a time when you are able to feel the pain of another and walk a mile in his or her shoes, so to speak. Since you are becoming more sensitive to your own emotional needs, you are also more sensitive and aware of the emotional needs of those around you. For this reason, you can become increasingly supportive of immediate family members. Dependency is a possibility. The Moon in the 1st can commonly be seen in the charts of those with a new baby or elderly parent to care for. Your ability to be more nurturing at this time draws those who are in need of your care and assistance. Whether you are male or female, strong involvement with a woman is especially common.

MOON IN THE 2ND HOUSE

The Moon in the 2nd house shows an emotional need for financial security. In order to feel comfortable during the next year, it is important for you to review your financial situation and formulate plans which will increase or insure economic stability. This financial plan should be capable of meeting your immediate and future material needs by addressing three major issues.

First, how dependent are you on others for your financial security? You may have been financially dependent on someone else in the past and at this point in your life, you could view your financial dependency as immature, unstable or undesirable.

Second, how stable is your present salary and what can you do to increase future earning power? Although a steady income would be reassuring at this time, it is more common to have a vacillating income. This is not the time to expect a straight, dependable salary. Most likely your income will fluctuate, but this could happen for several reasons. You might be self-employed or only work part-time when the need arises. Your schedule could be erratic or your earnings could be based on commissions or incentives. For some individuals, the Moon in the 2nd indicates a break in income, though usually this is not the case. Your overall income can actually change either positively or negatively, but it is more likely to increase if you actively focus on money-making ideas. The Moon in the 2nd house is generally a sign that materialistic attitudes are very strong and making more money is part of this trend. Even so, the need for money is usually more closely associated with an emotional need for financial security than greed or avarice.

Third, are you handling your present funds effectively in regards to spending practices, savings accounts and investments? Spending practices are changing and expenditures will go up or down depending on what you are purchasing and who is spending the money. If you are spending your own money, you are less likely to make frivolous purchases and more apt to budget, save or invest. You will tend to be more cautious and less speculative; with economic possibilities so varied, you would be wise to conserve money for rainy days and major purchases. Outside of investments, the one major purchase commonly associated with the Moon in the 2nd house is a home, particularly the initial house. This house is generally seen as an investment as well as a home, and a major step towards financial stability.

Since the interpretation of this planet in this house has centered so much on money, it is important to reiterate that the focus is not generally money needed to fund materialism, but money needed for financial security.

THE MOON IN THE 3RD HOUSE

The Moon in the 3rd house is a sign that emotional influences compete or unite with rational thoughts for decision-making power. The interplay between these two levels of thought, the conscious and the unconscious levels, is key to the pattern of growth. If you depend too much on the conscious level, you might be trying to suppress or analyze your feelings before they are even apparent. This will not work. Feelings just are, and must be appreciated as they exist before symbols emerge and they can be understood. On the other hand, you may feel more ruled by unconscious urges than clear-headed rationalism. Within this scenario, feelings overwhelm common sense and burst into the open when you least expect it. Freudian slips are possible despite your conscious efforts for restraint, and you inadvertently say things that were better left unsaid. Although these comments honestly reflect your true feelings, they go too far and reveal more than is in your best interest or the best interest of others.

The key to working with this placement is understanding the need for conscious and unconscious appreciation, integration, and consensus. This is not the time to depend totally on either conscious thoughts or feelings. It is the integration of the two levels of understanding which is most informative. Emotions will be the basis of learning, but once feelings have reached the surface, it is most insightful to understand them within the context of the total experience, both rational and emotional. For example, if you attend group therapy sessions during the year, you will focus on the psychological meaning of what is being said rather than the actual physical details. However, once the emotions have risen to the surface repeatedly, patterns of response will be intellectually understood.

The combination of an emotional planet in a rational house can also indicate that you get mixed messages. You may not be exactly sure when you are thinking and when you are responding from the gut level. You may even be of two minds, especially regarding a relationship. Your heart tells you one thing while your mind tells you another. Competition, rather than integration, is possible in those who do not work towards comprehension. Positively, this is a strong combination for those who yearn to integrate feelings and thoughts, and experience life as multidimensional yet whole. If you are a writer, artist, psychologist, counselor or holistic practitioner, the placement of the Moon in the 3rd house can indicate that you experience the conscious and unconscious mind as a unit working together. The artist can use creative projects to explore and understand emotional themes. If you are a counselor or involved in therapy yourself, you can use the conscious mind to decipher and interpret unconscious messages. Used wisely, mixed messages can lead to great insight.

You can become a more public personality, especially in your own neighborhood or area of expertise. For example, one individual with the Moon in the 3rd house of the solar return had been working for a large organization. He then started his own business and began pushing his own products. He was lifted from the obscurity of the company to a more public position. What you do and say will become more noticeable this year. Your demeanor and presentation are important. You can make a lasting impression on those who see or hear you in a public capacity. You may even be a role model for others. Therefore it is important that you maintain a good public image and reputation.

Neighborhood selection is important to those who are thinking of purchasing a home. If you are already settled, this is a good time to become more involved with your neighbors and community associations.

MOON IN THE 4TH HOUSE

A 4th house solar return Moon suggests an emphasis on two important areas: the physical home or domestic situation, and the emotional needs and perceptions of the individual. Usually these two major areas of concern parallel each other and changes manifesting in the home or home life are reflected in emotional changes the individual is experiencing internally. For example, if you do extensive remodeling of your home, it is likely you will also experience emotional disruption during the period of renovation. It can be argued that the emotional disruption is directly caused by the upheaval in the physical environment of the home; but, it is also possible that your emotional situation is changing rapidly and internal dissatisfaction actually caused the desire for renovation.

Perhaps a different example will illustrate this better. Whenever one moves into a new home, some amount of redecoration takes place. The priorities set at this time can be indicative of the individual's emotional climate. Those who are withdrawn or recovering from a difficult emotional trauma might decorate the master bedroom first. The master bedroom suite is probably the most private part of the home and is not always seen by a visitor. Yet it is viewed as the most restful room in the home, a protective inner sanctum that revitalizes energies. Establishing a priority of redoing this room first may reflect a desire to soothe injured feelings before all else. In this manner, internal emotional needs lead to external choices made in regard to the home.

You are apt to either change your place of residence or change something about your home when the Moon is in the 4th house of the solar return chart. You can purchase a new home and move, but it is just as likely that changes will occur within your existing home or domestic situation. For example, you could have a household member come or go during the year, perhaps a new roommate moving in, or a grown child moving out to a separate residence. Feeling "at home" becomes very important and you need to make changes in your home or lifestyle to attain this sense of comfort. Your aim will be to create a place of retreat reflecting your individual emotional style and needs. There are a number of different ways to do this and changes can be either physical or emotional, radical or minor.

Occasionally, domestic needs affect career goals. Interest in providing a comfortable home environment may become a career concentration. It is possible to begin a career in real estate, home design or interior decoration during the year. Security issues related to your home or future retirement may lead you to plan a career move or to change your job.

Emotional feelings are the second major area of concern with the Moon in the 4th. Your emotional needs are stronger and more urgent at this time, probably because you are coping with an element of change or disruption. You are very aware of the understanding and emotional support you receive from others, or the lack thereof. You want to feel a strong emotional connection to those around you, and need greater intimacy. If others cannot or will not match your needs, you withdraw emotionally and physically from disappointing situations. You tend to be more protective of others and can draw dependent relationships during the year if you feel the need to mother or be mothered. If grown children are leaving home, it is possible to experience the empty nest syndrome. New mothers may experience postpartum blues. These feelings are all related to a maturing emotional nature and changing circumstances. Focusing on your own emotional needs at this time will help you to make the necessary adjustments.

Changes in your emotional circumstances can be more relative than actual. Your situation stays the same, but suddenly you have realizations which open your eyes to new possibilities and problems. Intuitively, you now sense much more than you are told. This ability can give you a wealth of new knowledge and insight, though you might be reluctant to trust it at first. What you sense intuitively may or may not be different from what you are told verbally. If there is a discrepancy, you'll feel torn in two directions and not know what to believe.

Extreme sensitivity to issues related to the home is common, and what actually transpires in the domestic environment, physically and/or emotionally, tends to affect you deeply. Aspects to the Moon or other planets in the 4th house imply either increased or diminished physical and emotional satisfaction. If you are in a very difficult situation and your home life is disrupted, it is possible to feel homeless with this placement, but for those who work with their feelings, there awaits a new sense of roots.

MOON IN THE 5TH HOUSE

The start of a new romantic relationship is the most common event associated with the Moon in this house. This new relationship can be viewed as a "first" in one way or another. If you are very young, it may be your very first relationship or first meaningful relationship. If you are a little older, it can indicate the year you begin having sex, especially if you are emotionally involved with one particular person. It might be the first time you live with someone or begin dating the person you are going to marry. If you are older and more established in these areas, it may be the first time in a long time that you are emotionally involved with a new person. Regardless of your marital situation, this can be a year when you feel an emotional connection to a new person, and usually this connection is very compelling.

Unlike the Sun in the 5th house, the Moon is not necessarily associated with sexual attraction. The Sun seems to be more suggestive of a sexual affair. In fact, with the Sun in the 5th, the relationship may simply be a sexual fling with limited emotional contact. But the Moon is more closely associated with emotional needs in a relationship, and therefore the contact may or may not be sexual, though many times sexual involvement occurs. It is the desire for emotional sharing and expression that seems to be the main focus and motivation for the relationship. Then, as the relationship deepens and intimacy increases, sexual involvement develops. The need for emotional expression is probably the key to understanding the interpretation of the Moon in this house and romance is a wonderful means for expression. Though the give and take of a relationship is just one way to handle feelings, it is a way many people seem to choose.

If you are involved in a relationship such as this, your feelings will be very intense. You are more apt to express yourself emotionally, and consequently it will be more difficult to hide what you are feeling. This may not be a good time to keep a clandestine relationship secret. You will be subject to emotional swings. At times you will be elated, while at other times you might feel depressed. Your feelings may depend on how well the relationship is going and whether or not you can see each other. An element of "need" is many times associated with these relationships and dependency is a problem. The individual you are involved with could need your help in one way or another. The situation need not be crucial, but you will have the opportunity to express nurturing and protective feelings. If you respond in a mothering or overprotective way, the relationship may involve dependency issues. On the other hand, you may be the one who becomes dependent.

Your feelings are not necessarily stronger and deeper this year, but you are more apt to express them in an obvious way. This may include openly stating what you need or want. Being direct is probably the best way to express your feelings, but it is also common to seek secondary or alternative ways. Finding satisfying means for emotional self-expression is one of the tasks associated with this placement and may contribute to the romantic nature of the year. You may choose an artistic or dramatic means for emotional expression. You could, however, choose creative endeavors such as painting, writing, or acting. Or you might want to participate in intense situations which exemplify the drama of human life, such as being an emergency room nurse, or soup kitchen volunteer.

Your strong need for emotional expression can lead you to intense personal situations. You may have a baby instead of a love affair. If you are aware of your emotional needs, you can consciously choose an outlet for this energy. There is no limit to the manifestations you can pick so long as the situation allows for emotional involvement and expression. If you are frustrated by your present circumstances and the people around you are unable to satisfy your emotional need for expression, look elsewhere. You need situations which allow for some external display of emotions. For a few, this may involve living life on the edge of an emotional crisis.

One final avenue for emotional expression can be intense involvement with children. You may be especially concerned with their emotional well-being or creative abilities. If you have children of your own, this can be a good time to reach out to them. Focus on understanding their needs. Your children may be more emotional this year, especially if your relationship with them is changing or changes are occurring in the home environment. They may require extra love and attention if this is the case. The difficulty with the Moon in the 5th is that a parent might view his or her children as an extension of his or her own personality. As a parent you can have certain needs that you assume only your children can fill, and this places unnecessary pressure on them.

Aspects to the Moon in the 5th house can suggest avenues for emotional self-expression or possible inhibitions and blockages. Aspects to the 2nd house imply that your mode of self-expression can be either financially lucrative or expensive to maintain. Expenses related to children may increase. Traditional values associated with the 2nd house may or may not be compatible with your romantic choices. A conflict is generally the case if you are married and involved in an extramarital affair. Aspects from the 5th to the 8th house can also suggest financial concerns. If your children are starting college, you might assume a large loan to cover their education. A sexual issue or problem could affect any relationship you are involved in. The Moon in the 5th aspected by planets in the 11th symbolizes a freedom-closeness dilemma involving romantic involvements. "Are we serious lovers or are we just friends?" typifies this dilemma.

MOON IN THE 6TH HOUSE

The essence of the Moon's nature is change and fluctuation. Its presence in the 6th house of the solar return can denote changes in working conditions and/or health and dietary habits.

Job changes are usually minor. The Moon is more closely associated with adjustments rather than major upsets. Some appropriate examples of these minor changes might be: you change departments or offices within your company without changing your job per se; you drop your full-time hours to become a part-time employee; if you are self-employed, you may hire an assistant to help you keep up with the demands of your growing business; or if new policies and procedures are instituted at your office, your daily job routine changes. The adjustments that occur may or may not be within your control and may or may not be beneficial.

With the Moon in the 6th, you are more likely to make decisions that either directly or indirectly cause job changes. You can consciously decide to make the changes yourself, or make changes in other areas of your life which eventually cause repercussions in your job. As an example of this, you could decide to move out of the city. Since you will now live farther away from your office, you convert a room in your house to a home office and do more work at home. It's possible that changes occurring in your job are truly out of your control. Your company may merge with another, reorganizing the offices and moving you to another department. All changes tend to be minor, even those that are beyond your control. Except in rare instances, it is unlikely that you would quit your job or get fired during this year; however, job security may be an issue. Even if the Moon is heavily aspected in the 6th house, it is more common to experience daily routine changes that are complicated and stressful rather than major career transitions.

There is a tendency to become more emotionally involved with co-workers when the Moon is in the 6th house, and if you are in a helping profession and/or have daily contact with the public, you may become more involved with those you help. This interaction includes a greater sensitivity to the needs of others while on the job. Because of your dealings with the public and co-workers, you are more apt to develop a reputation. This reputation is dependent on the quality of your work and therefore may be either good or bad.

The Moon in the 6th also implies health changes. These changes are generally minor and can be caused by physical changes in the body or adjustments in health habits or diet. Body changes are usually the result of natural biological cycles and processes. Included here are those changes resulting from puberty, pregnancy, breast feeding, aging, physical fitness, and hormonal balance or imbalance. Weight gain or loss is also possible. It is likely that you will be more mindful of what you eat even if you are not eating well. The body is more sensitive to the way food and exercise (or the lack thereof) affects you. This is not a good time to go on a strict health regimen. It is a much better time to learn health from the inside out. Be aware of what foods and activities make you feel better and which ones actually hurt your health or drain your energy. Develop a health consciousness. This year conscious changes in health habits are more likely to be made because of greater awareness of how unconscious habits affect health. Make your changes gradually and incorporate them into your new lifestyle permanently.

Any illnesses you might have during the year can relate to emotional habits you refuse to change. It is very easy for psychological problems to manifest as physical illnesses. A healthy diet is not enough. You must also have a healthy emotional life, which includes fulfilling emotional involvements and psychologically healthy environments. Relationships and work habits can affect your health, especially if you experience a lot of stress, frustration, or anger in these areas. Look at your most intimate relationships. Are these relationships assets or liabilities, fulfilling or draining? Even if you find these relationships debilitating, you have recourse. Counseling can help you improve your situation or meditation can help you cope with the stress. It is also important to evaluate your working conditions and the effect they might have on your health. You should be able to make changes here also. If you are aware of your health needs and take corrective action, you do not have to be sick. The more you ignore the situation, the more difficult it may become. See a psychological counselor or holistic health practitioner if you should feel the need. It is a good time to acquire good emotional habits.

MOON IN THE 7TH HOUSE

The Moon in the 7th house of the solar return shows the probability of being involved in a nurturing relationship. This relationship does not have to be an intimate one, though this is a possibility. The caretaker quality is pronounced even in business relationships. You may be caring for family members or you may donate your extra time and money to a needy family that you know. If you are a secretary, you may be very involved with your boss's personal needs and comforts rather than clerical duties. If you are a physician seeing patients, you will be very concerned with your patients as individuals.

Personal involvement and concern is characteristic of this placement. It is very likely that you will personally relate on a regular basis to the individual you are helping. And unless the relationship is also an intimate one, it is common for the relationship and the nurturing activities to appear one-sided. Non-intimate relationships associated with the Moon in the 7th house are not equal. One person has more knowledge, expertise, status or responsibility than the other; one has more power and control than the other; one is cared for and nurtured while the other does the nurturing; one shares his or her feelings and the other just listens and/or helps. It is very common to be involved in professional or informal counseling relationships during the year. The individuals tend to be unequal and one person gives while the other receives.

Intimate relationships have a similar, yet slightly different, manifestation. Dependency issues are also common in these relationships and the individuals involved tend to assume polarized roles during the year. Examples include caretaker roles, stronger versus weaker or traditional male-female roles. Existing relationships change in some way and many times this is the change that occurs. Even if you and your partner have established a balanced relationship over a long period of time, it is usually the case that during this year, one, more than the other, needs to be supported emotionally, economically, or physically. The more dependent individual in the pair finds it difficult to make decisions, be assertive, or handle daily problems. The more dominant individual is usually in control and becomes responsible for the couple's future and welfare. Occasionally this is caused by illness, but it is more likely to result from subtle shifts in power within the relationship, or a greater need for understanding and support. Long-range goals, education, pregnancy and child rearing may contribute to these power shifts. In very strained and difficult relationships, the balance of power is heavily weighted in one direction. The weaker individual finds it hard to attain a sense of individuality and yet is afraid to leave the relationship entirely. In very balanced intimate relationships, mutual nurturing is possible.

You may meet someone and become romantically involved during the year. You need greater emotional intimacy. Ideally, you will be able to establish a pattern of mutual support. But the Moon in the 7th house does not guarantee that your relationships will be naturally fulfilling. If you are involved with a partner who is capable and willing to match your need for closeness and sharing, then it is likely that your relationship will deepen and grow. However, if your relationship depends on your ability to care for the other's needs, and your partner is unable or unwilling to reciprocate now or in the future, then your relationship will be emotionally draining and difficult. Expect your moods to change with the positive and negative shifts in this relationship. Since change and fluctuation will be such a major factor in your relationships during this year, moodiness can be a problem until you learn to handle the changes.

You will tend to be more emotional than rational this year and you will follow your heart more than your head. This is because you may be more interested in emotional fulfillment than intellectual analysis of your situation. Check the placements for Mercury, Pluto and Venus to better understand the Moon in the 7th house. If you are truly in love with someone and the relationship is good, you will be able to support each other and overcome any obstacles. Your relationship will grow more intimate. But strained or blocked relationships will only become more difficult. It is the emotional connection, whether good or bad, that draws you to someone; unfortunately, this need for intensity and lack of objective thinking may enable you to remain in a nonproductive relationship with unrealistic expectations for the future.

The Moon in the 7th house is also a sign that unconscious complexes can complicate your present relationship. Former relationships which reinforced negative patterns of relating and left you feeling wounded may have left psychological scars which must be faced and dealt with now before greater intimacy can develop. It is essential that you seek to understand these complexes. Irrational fears, obsessions, possessiveness, jealousy, and lack of trust are just a few of the forms these complexes might take. This is an excellent time to see a marriage counselor or attend a marriage encounter. If your relationship is basically combative and neither of you is interested in deepening your commitment to each other, then emotionally difficult scenes will be generated by your feelings of hurt and rejection, both in this relationship and others. In this case, individual counseling is more appropriate.

MOON IN THE 8TH HOUSE

The 8th house is the house of acute awareness of emotional forces and psychological powers. Nearly all of this awareness will result from a newfound ability to spontaneously perceive information on these subtle levels rather than from a rational analysis of any situation. It is the intense awareness of unconscious emotional and psychological issues that forces conscious insight into human behavior. These realizations can be upsetting and you may end up seeing or knowing more than you care to see or know. Acute awareness is also associated with the Sun in the solar return 8th house, but with the Moon here you are more apt to be emotionally involved with the people and issues concerned. It is likely that someone in your immediate family or circle of friends needs counseling. During the year you grow more aware of your own emotional needs and psychological forces and you also learn to recognize these influences in others.

It is very likely that you will study some psychology during the year, but you may become aware of psychological influences naturally without any education. Initial insights can come from associating with mentally disturbed people. The intensity of their psychological abnormality is fascinating, yet frightening. It is common to have an 8th house solar return Moon if you are a professional therapist or astrologer just starting to counsel others. Disturbed individuals need not play an important role in your life and may be mere acquaintances. They are not the real focus of your realizations, they are merely the catalyst.

You need to become more aware of the psychological influences affecting you and your immediate relationships. Intense realizations associated with bizarre personalities will give way to insights into subtle manipulations and power plays. Eventually, psychological realizations can occur during the simplest interactions with others. If you are caught in an emotional power struggle, it will become more obvious how you and/or others manipulate rather than state clearly what you need. Psychological complexes and hurt feelings complicate relationships and make it more difficult to discuss problems rationally. Unconscious feelings may undermine conscious decisions. Ambivalence is common. Obsessive-compulsive behavior, jealousy, guilt, unexplained anger and helplessness are frequently clues that something is wrong.

Recognizing and understanding your own psychological idiosyncrasies will give you great power over your own life. Insights into your defensive reactions will help you overcome barriers to intimacy. Increased intimacy is especially important since superficial relationships and superficial discussions will not suit you this year. You need to be able to talk at length and in depth about what you are feeling and perceiving. It is during these discussions that initial realizations are triggered and later confirmed. It is also during these discussions that you formulate and express new behavioral patterns that sidestep psychological inadequacies.

Emotional situations tend to be more complicated and intense. Issues are no longer black or white, good or bad. Decisions are less straightforward, especially if you feel responsible for the well-being of others. For example, if you have children and were offered a transfer to another state, you would not make this decision without careful consideration. On one hand, the move may improve your career possibilities, but on the other hand, the move could have a great impact on your children. You must consider the emotional or psychological ramifications of your decisions since they are more likely to affect others.

If partners or family members are moody or unstable, you may have to think before you speak if you want to avoid upsetting responses. You may be in a delicate position requiring great diplomacy and tact. Others can interpret what you say incorrectly or overreact. You will be dealing with intense feelings and may expect others to involve you in very emotional circumstances. If you counsel others professionally, it will be easier for you to control these situations. If these situations are appearing in your personal life, you can deal with them effectively on your own or seek advice.

Increased intuitive accuracy will encourage your reliance on this mode of perception. Psychic and intuitive realizations grow stronger, but may seem less controllable. Even if you do not consider yourself psychic, this is a good time to trust your hunches, but if you are psychologically stressed, you may be less able to distinguish true psychic impulses from fears and anxieties. If this is the case, you should reserve judgment until your powers of discrimination grow stronger.

The Moon in the 8th house can show changing shared resources or economic dependency. If you share funds with your spouse or some other person, the amount of money you receive will change, either positively or negatively. This change may be reflected in the aspects to the Moon. On the other hand, your own personal resources are more likely to decrease in comparison. The reduction need not be much, but you are more likely to feel like a dependent rather than an economic equal. There could be very valid reasons for this. For example, you may be on pregnancy leave or attending school. Financial power struggles are symbolized by the Moon in the 8th opposed to a planet in the 2nd house. You and your partner or someone else may fight over how to handle money or pay bills. These arguments might be very emotional and further complicated by unresolved psychological issues. Clarity and a willingness to discuss needs will help you to reach positive compromises. The Moon in the 8th house also indicates a change in indebtedness. Debts either go up or down, but they rarely stay the same.

MOON IN THE 9TH HOUSE

The 9th house is the house of beliefs: religious, philosophical and mundane. The Moon placed in this house suggests a good time to reevaluate all beliefs in light of practical applications to daily situations. We think of the Moon as representative of common everyday experiences and domestic influences; therefore, your beliefs must now pass the test of useful application. Personal philosophy should help you cope with everyday situations and issues facing you. Higher knowledge and beliefs must be consistent with what is most practical and efficient on a mundane level. There might be times during the year when you feel you cannot reconcile higher knowledge with practical experience. What you believe does not help you with the decisions you must make. Discrepancies like these point out useless beliefs and the need for review. Emotional feedback from everyday situations and interactions with others will help you to find out which beliefs are working on a mundane level, and which ones leave you feeling vulnerable to life's pressures.

Your old beliefs may be outdated. Beliefs can grow stagnant from a lack of review, and you must keep your philosophy current and consistent with your pattern of personal growth. Beliefs are sometimes carried over from someone else's system. Don't accept higher knowledge, religious or philosophical, blindly; what others espouse may or may not fit your needs. You will find that some enlightened thoughts are not helpful when it comes to functioning on the mundane, everyday level. They do not translate into practical rules to live by. Your philosophy must apply to your everyday circumstances. Review beliefs and personalize those which have meaning within the context of your own life.

You could be asked to explain your beliefs in public, or teach, or write a book. In some way, you might be called upon to explain your beliefs. This is a good time to practice what you preach or make adjustments in the sermon. This is the key to utilizing the review process. Practice helps you formulate a philosophy that fits your everyday life. You retain what works and discard what does not. Through this testing process, you can develop a philosophy that works for you spiritually, emotionally and physically.

Because the Moon is in the 9th house, what you give and get out of relationships will be very important. Our beliefs set the tone for our interactions with others, and our responses to daily situations. For example, a common Christian belief is that one should turn the other cheek if offended, assaulted, etc. While this might be good advice for some situations, it may not be helpful all the time or in your particular situation. If you are being repeatedly abused, verbally or physically, without resistance, it might be time to develop a new tactic. A religion/philosophy which encourages you to remain in an abusive situation is questionable and should be reevaluated.

Your beliefs about relationships regulate the demands and expectations you have in a relationship. Your life is at least partially structured by your beliefs. If you think that all men are users, then you will tend to draw individuals who confirm your expectations. This is a good time to review your beliefs about others and relationships. It is likely that some beliefs will be changed and modified as you begin to work towards a more successful philosophy.

You may be involved in a legal matter during the year, and consequently become more familiar with the legal system and the judiciary process. You may consult a lawyer or you might be involved in a court case as a plaintiff, defendant, witness or jury member. You may be interested enough to follow a particularly fascinating court case. Legal matters are not necessarily strongly emphasized. The most common manifestation is seeing a lawyer during the year.

You may return to school or take a course during the coming year. Topics may be academic, recreational or artistic. On the other hand, you could be teaching rather than studying. Teaching may help you ground your spiritual thoughts. Certain topics require examples and applications from real life to be understood.

Travel is also a possibility with the Moon in the 9th house. Usually the more planets in the 9th house, the more likely you are to travel. Travel can be to a foreign country but you may only do extensive traveling within this country. It is also possible that you commute regularly and over a fair distance for your job or for other reasons. Increased travel time may occur without foreign travel per se.

MOON IN THE 10TH HOUSE

The Moon in the 10th house suggests changes in career, or professional tasks you are asked to perform. While the Moon in the 6th house can indicate minor changes in working conditions, the Moon in the 10th house generally implies changes that are far more significant, either in the immediate future or several years down the road. Minor changes can occur and may involve shifting positions or departments within the same company, but it is more likely that this shift includes important changes in your job tasks. For example, a secretary accustomed to working on a typewriter was asked to do secretarial work on a new computer. Eventually, this individual went on to a career in word-processing because of this experience.

Changes occurring during the year can be very beneficial. This is a good time to focus on career development and take advantage of professional opportunities as they arise. Many times your reputation plays a key role in the professional events of the year. Everything you do at work will be more public this year. If you perform well, you will receive the recognition you deserve; but if you perform poorly, your mistakes will be very noticeable.

Job security can be an important issue and you may feel that your position is threatened in some way. The company you work for could be having financial difficulties, or conditions within the company might seem unstable. The Moon is less apt to indicate that you lose your job or are laid-off; serious job changes are more likely to be suggested by Saturn or Uranus in the 10th house. However, firings and layoffs can and have occurred while the Moon was in the 10th house. Usually the individuals involved were controversial figures who were unpopular, and had acquired a negative reputation. As a rule, most changes are directly or indirectly within your control. Even those who lose their jobs play a role in their own misfortune.

The Moon can indicate public recognition as well as recognition within the company. You can have more contact with the public than previously. You could move to a position involving public service, relations or communication. For some individuals, the Moon in the 10th house indicates notoriety and a public reputation. The publicity can be either good or bad. This is a good time to focus on your dealings with the public and to use the media positively. If you are a politician, you can develop a following. If you are a salesperson, you can key in on the wants and needs of your customers. Emotional intuitiveness can serve you well in your dealings with the public at large.

MOON IN THE 11TH HOUSE

The Moon in the 11th house suggests that you have a dream or goal you are working toward. This goal should be personally fulfilling and rewarding, when and if completed. Major tasks started during this year generally involve personal commitment to a project, idea or belief. You should be your own motivator, and many times you will be working alone. The goal need not necessarily be humanitarian or idealistic. Teenagers trying to pick a college or career can have this placement. They gather the pertinent information and make choices according to their needs.

You might have to work towards your goal despite much opposition and lack of assistance from family members. For example, a woman with the Moon in the 11th started her own cottage industry despite her husband's insecurity and messages of doom. Opposition to your goal can be a blessing in disguise if the tension increases your conviction and dedication to the task at hand. It can be a motivating force that pushes you onward. This is a good time to accomplish something that is truly your own. The Moon in the 11th implies personal goals. Therefore, it is very important that you reassess your present and future goals to determine if they are truly the product of your own needs and desires. Do they have great personal value to you as an individual, and will they be emotionally fulfilling when completed?

You may draw closer to friends during this time while becoming more detached from your family. This is especially true if family members object to your future goals or present situation. The Moon in the 11th suggests that you develop closer bonds in what are normally more detached relationships. Friends become family and you are more apt to confide in them. They, in turn, are more likely to help you with your project than your own family. If you are at an age when peer groups are especially important, you will depend heavily on the opinions and support of friends. There could be a logical reason for this. For example, if you are selecting a college or job, other teenagers should know a lot about schools and work programs. Information acquired from other students or through the grapevine is most helpful. In this case, the dependency and close contact with friends are important to the task at hand.

New patterns of closeness and independence evolve during this year and some vacillation may occur in all relationships. The basic interpretation for the Moon does not blend easily with that of the 11th house. The Moon is emotional and dependency-oriented while the 11th house is detached and independent. The lack of common ground can suggest a freedom-closeness conflict. You may be at a critical time period for your emotional maturation when you are about to take a big step towards greater independence. You could be planning to go away to school, live on your own, become self-employed or self-supporting. As much as you welcome the move towards greater independence, you will also be anxious, hence the frequent revisions in your plans.

Your sense of attachment to others will also fluctuate and you may experience some relationship issues. You may not be able to depend on significant others in your life for one reason or another. They may not be dependable or supportive, or you may not want their assistance. It is difficult to actively seek out nurturing situations while at the same time struggling to establish your independence. You can experience this freedom-closeness dilemma as relationship conflicts which seem to alternate between restriction and abandonment. Reassuring others of your love can ease these growing pains. Realize that if you appear unpredictable, others will feel threatened by your need for independence. Or this struggle might be completely reversed. You may be moving towards greater intimacy. If you have been on your own for a long time and you are now romantically involved, you could fear giving up your independence and committing to a permanent relationship. The dilemma is the same, only the situation is changed.

This placement is associated with all kinds of groups, but group meetings and activities will tend to be emotionally charged. Positively, you might be very concerned with an injustice or cause. The emotional impact of the group's efforts is a motivating force. Negatively, conflicts break out among members. Emotional and psychological factors within the group tend to complicate gatherings. This is an excellent time to join a therapy or self-help group. The Moon suggests emotional attachments within a detached setting, and you can spend a lot of time discussing your problems with your friends or group members. Whatever the situation, the goal is to share feelings. The give and take needed in relationships this year works well within support group situations where camaraderie occurs in a somewhat detached setting. In this type of group setting, you are able to experience strong emotions without being weighed down by emotional responsibility, a good combination for the individual with a freedom-closeness dilemma.

MOON IN THE 12TH HOUSE

You may not be very open with your feelings when the Moon is in the 12th house. This can be a time of intense emotions, but commonly they lack a clear avenue for expression. For one reason or another, you find it difficult to reveal what you are feeling or sensing intuitively. Even when you do make a revelation, you suspect you are misunderstood or ignored. Others may seem distant or detached, leaving you to cope alone with pressing emotional issues; or you might not have someone immediately available who truly understands what you are feeling. The Moon in the 12th house is common in the charts of those who have recently changed their environment and entered a new situation where they have not yet had the opportunity to foster strong emotional connections to the people around them. The perception of emotional loneliness may or may not be accurate; you could simply be more reluctant to disclose your emotions though others are willing to listen.

But there are certain situations associated with the Moon in the 12th house which typically warrant emotional restraint. If you are involved in a clandestine affair, the person you are seeing may not want to get involved emotionally. If you are dealing with the sick, disabled or dying, you might hide your feelings so as not to convey the full extent of their injury or illness. If you think that someone has purposefully tried to hurt you, you would not want him or her to see if they have succeeded. Consequently, this can be a time when you keep emotional secrets because of the anticipated reaction to their revelation.

The reasons for emotional restraint are very important. You must ascertain whether or not your situation truly prohibits emotional expression or if your own fears are limiting their expression. Even if your present relationships are restrictive, this does not mean that you cannot or should not find others to confide in. You can cultivate new friendships that allow a greater opportunity for expression. Learning to express your feelings quietly to those with whom you are most closely involved will be one of the tasks for this year.

You may be socially withdrawn and retiring, preferring to be left alone, withdrawing from the public eye. You might enjoy quiet moments at home with your family, friends or loved ones instead of social evenings out. You need extra time, and perhaps privacy, for your own interests. If you are normally an outgoing, active person, this may seem like a time of seclusion. If you are romantically involved with someone, you will want to be alone, or may have to be alone. The individual you are involved with might not want to be seen in public.

The Moon in the 12th house indicates that the emotional nature is more sensitive to those in need. This can be a year in which a lot of time is spent helping, nurturing or caring for others, especially if there is someone close to you who has a health problem. This Moon placement can suggest that someone close to you, and possibly a family member, is sick or requires hospitalization. If no one in your family is ill, you will still care for someone who is in need. You can be a Good Samaritan who willingly gives time and energy to assist others. The danger with this placement is the tendency to get involved with those who refuse to help themselves, e.g., the alcoholic, drug user, unhappily married person, or counseling candidate. You, yourself, could play victim. If this is the case, seek professional guidance in these matters.

If the Moon will cross the Ascendant by progression (approximately one degree per month) during the solar return year, this can indicate a time when emotions are released or matters emerge more into the open.

CHAPTER FOUR

MERCURY IN THE SOLAR RETURN CHART

Introduction

Mercury has two basic interpretations in the solar return chart: it symbolizes both your mental condition during the coming year and what you are thinking about. It is generally not indicative of your sister, brother or relatives. It won't reflect correspondence or mail and it is not a good indicator of travel. However, a wealth of information can be obtained by understanding Mercury's contribution to the solar return's interpretation.

Indicators of Mental Conditioning

The first task involved in understanding Mercury's interpretation is to evaluate the indicators for mental conditioning and the possible sources of tension. Modal preponderances illustrate different thought processes and therefore are related to different sources of stress in the solar return chart.

CARDINAL PREPONDERANCE

A cardinal preponderance can show a shortened attention span and a variety of interests, especially if Mercury is also in a cardinal sign. Overscheduling may occur. If the individual is trying to do too many things at once, he or she can be stressed by a very demanding schedule. This individual needs to spend a few moments each morning organizing the day's events and writing down important things to remember. Stress can be reduced by setting aside time to simplify and organize the day ahead.

FIXED PREPONDERANCE

Many fixed planets can indicate stress resulting from a fixation on one or more issues. Thought patterns may be very focused or even obsessive, especially if Mercury is in a fixed sign. There can be a tendency to see things one way and one way only. Individuals fixated on a particular relationship, situation, or conflict need daily time off for stress reduction techniques. They also need to search actively for alternative solutions to problems or alternative ways to get what they want. This ability to focus can be very productive for those individuals who are intent on finishing a major project, but it can also cause great frustration and/or mental blockages. Breaking concentration (at least once a day) with a right-brain activity can allow new thought patterns to emerge. Productivity may be increased while stress is decreased by these creative periods.

MUTABLE PREPONDERANCE

Mutable preponderances can show a lack of focus and organization. While the cardinal preponderance indicates a scattered focus and the fixed preponderance indicates a concentrated focus, the mutable preponderance can be the inability to focus clearly on a particular issue or situation. Normal patterns for organizing thoughts are sometimes disrupted by changes. Individuals temporarily lack good discriminatory powers, becoming indecisive and easily swayed by others, especially if Mercury is also in a mutable sign. Nervousness and stress result from feeling overwhelmed by external influences. These individuals need time alone each day to reflect on what is really important. They are very open to advice, information and new ideas, but they need to retreat from these influences long enough to assimilate and evaluate what they are experiencing. They need to weigh information and develop new criteria which will help them make good decisions. This is especially important if they are in the process of reorganizing their lives after major changes. In this way, stress resulting from worry, indecisiveness and change may be avoided or minimized.

AIR OR WATER PREPONDERANCES OR LACKS

The number of air or water planets in the solar return chart can also help you to understand the mental conditioning and the origins of stress. A chart with a preponderance of air planets may indicate a greater attention to the spoken and written word. Stress results from the need to remember numerous details and, sometimes, tedious information. This preponderance is sometimes seen in those who are in school, trying to pass a professional examination (such as the bar exam), mastering a new job, or doing research. A tendency to think rather than feel is likely, especially when the air preponderance is reinforced by a lack of water.

Emotions may be blocked or devalued. If this is true, personal feelings might not be considered when making decisions. The individual can rely on conscious information while ignoring intuitive impressions and emotional needs, and totally underestimating unconscious forces. This is a time when one could make an intellectual decision which is emotionally difficult to live with. Conscious/unconscious inconsistencies of this nature need to be investigated and corrected. Artistic or right-brain activities can help you to release tension and get in touch with emotional material which is not easily accessible.

On the other hand, a chart with no air planets and a preponderance of water symbolizes a year in which rational thoughts are overwhelmed by emotional concerns. You may be ruled by your heart and not by your head. Unconscious reactions to events are more likely than clear logical responses. This is a time when individuals go to extraordinary lengths against all odds because they truly believe that they can make a difference. Many times they do. The immediacy of emotional issues far outweighs other concerns. For those involved in a truly loving relationship, love may conquer all, but for those who are mentally unstable, this can be a very difficult time. It's great to be overwhelmed by love, but overpowering obsessions, phobias and anxiety are considerably less desirable. The preponderance of water can indicate that unconscious behavior is more prominent than conscious behavior, making the individual feel out of control. One's perception of reality can be distorted by projections and a lack of objectivity.

Regardless of your circumstances, stress can result from the lack of consistency between the conscious and the unconscious levels. Defying all reason creates stress. So does avoiding reality. Stress can be alleviated through positive nurturing experiences. Channel emotional energy into relationships capable of greater intimacy. This is also an excellent time for insight, since emotional issues will be so pressing.

A balanced number of air and water planets can indicate that conscious thoughts and unconscious feelings are well integrated and the individual is able to assimilate information from a variety of sources. This is most likely to occur in those individuals who have actively worked on this integration. Or possibly, it might indicate that the two elements symbolize competing psychological forces. Balanced elements are not always associated with integrated psyches. This tendency towards a lack of integration is sometimes suggested by the solar return chart. Moon and/or Pluto placements in the 3rd house while Mercury is in a water sign/house and heavily aspected is one way this competition might be symbolized. Another possibility is two separate solar return planetary groupings, one symbolizing conscious thoughts while a second symbolizes unconscious emotions. Individuals with placements similar to these can experience events as both dual and integrated. Discrepancies can occur. What they are told may differ from what they intuitively know; what they think may be different from what they feel. Inconsistencies of this nature suggest the need for better integration of conscious and unconscious forces. But these placements can also be found in those who are very aware of psychological forces and who can discriminate between conscious and unconscious information.

The stress interpretations presented here are meant to serve only as guidelines because many contraindications might exist in the solar return chart. Always interpret the whole chart and not any one part alone. Certainly Mercury's house placement and aspects should be considered since they describe what the individual is most likely to be thinking about and what personality traits affect the thought processes.

Planets Aspecting Mercury

Aspects to Mercury in the solar return chart can also be indicative of mental conditioning because they suggest that rational thinking is either enhanced or impeded by other personality factors or interests. For example, strong 11th house placements aspecting Mercury can indicate goals which motivate the individual to return to school or learn new information, but these placements can also suggest that goals are unattainable at this time because of a lack of education. Mercury conjunct Neptune in the 1st house square to Saturn in the 10th house can show the personal desire to apply spiritual concepts to business practice. On the other hand, they can also suggest frustration with present career choices and a lack of direction for the future. Various personality preferences and psychological factors can limit your ability to think clearly by dividing your attention and creating mental stress. Planets aspecting Mercury in the solar return chart denote how one's mental capabilities might be impeded or enhanced.

MERCURY-PLUTO ASPECTS

Pluto aspecting Mercury in the solar return chart may indicate that your conscious mind is very aware of unconscious material and psychological complexes. This awareness may originate from naturally occurring insights into human behavior or educational pursuits. You are better able to perceive what is unspoken or hidden. Motivations will be clearer to you even when they are not stated. Manipulations and psychological games will also be more obvious. Most likely this awareness will not be one-sided. You will be as aware of your own unconscious nature as you are of psychological complexes in others.

If you are involved in repetitious verbal battles over ideology, prejudice or intolerance, realize that these confrontations are related to your own psychological tendencies. They are not the sole product of other minds. Being very aware of these psychological forces can be stressful, especially if you know more than you are capable of handling. This is an excellent time for counseling, should you feel the need. The implication here is that knowledge is power, and specifically in this case, it is knowledge about the unconscious mind which conveys power to those who are aware of it and able to gain insights from it. This information is as valuable as intellectual facts gained from school. The more you understand about the unconscious, the more you are able to control your own impulses or resist the manipulative behavior of others. Although initially this increased psychological awareness may be stressful in itself, the understanding and control you achieve in the end can actually lead to stress reduction.

MERCURY-NEPTUNE ASPECTS

Neptune in aspect to Mercury in the solar return chart suggests working with the less clearly defined psychological drives such as compassion, creativity, and spirituality. Your sensitivity to subtle emotional connections among all people increases your concern for certain individuals in particular and humanity as a whole. You are able to acquire information through intuitive and psychic insight. Dealing with these subtle themes can lead to some uncertainty and confusion in the thinking process. Increased intuitive or psychic awareness can precede the ability to weigh this information for its accuracy. It is sometimes difficult to discriminate between what is really psychic and what is more closely akin to worry, fear or false hope. It is also difficult to find practical applications for idealistic concepts and inspirations which are represented by this combination. Therefore, stress can sometimes be associated with the more spiritual manifestations of Neptune-Mercury aspects.

Important factual information that you receive during the year may be partial, inaccurate or vague. Sometimes secrecy and deception play a role. Your normal points of reference for evaluating information may be changing. Without adequate facts, you may be left hanging for most of the year. It may be impossible for you to make a decision at this time or to evaluate your circumstances. You may be easily confused or misled, especially if the information you receive is inconsistent or incomplete. In older individuals, confusion may actually be senility. Neptune's most negative interpretation is a loss of mental capabilities through drug and alcohol use or abuse. This tends to be an uncommon manifestation that is more closely associated with an individual pattern of consistent negative behavior. Most people will not fall into this trap. They trust in the Universe even during times of uncertainty. Their focus is on compassion rather than a search for the truth. They understand that in the end, all will be known.

MERCURY-URANUS ASPECTS

Uranus-Mercury aspects suggest that you are open to new ideas which may take the form of new information you are learning or new concepts you are developing yourself. These aspects can show great creativity since they imply that the individual is able to approach problems from many different angles and is not locked into one structured way of thinking. Use this time to be innovative and original. Brainstorm with others. But, because your mind is somewhat unstructured, your ability to think clearly may be interrupted by erratic impulses and an inability to concentrate over any length of time. New information may be more exciting than reorganizing what you already know. If you must work on a major project that requires sustained mental energy, take frequent breaks.

There is the possibility that you will subject yourself to psychologically stressful situations during the year. Increased nervousness, anxiety and irrational thinking may be directly related to these stressful situations. Question the necessity for excessive tension in your life. If possible, withdraw from those situations that tax your mental and physical health. You may want to practice relaxation techniques and avoid stimulants. Work on calming and nurturing your nervous system.

MERCURY-SATURN ASPECTS

Saturn aspecting Mercury suggests a more serious and structured perception of reality. Life is organized in such a way that decisions and changes have serious consequences. Choices may be studied in depth before decisions are made, and it is likely that at least one major decision will be made during the year. Sometimes this decision is made under stress and usually it involves great responsibility. For these reasons, the individual tends to be conservative. He or she is looking for changes that produce greater stability, not chaos. Consequently, this is a good time to work on the completion of a major project. You have the ability to channel mental energy into constructive pursuits, even to the point of forcing yourself to finish a difficult and boring task.

Structured learning is associated with Saturn-Mercury aspects and some individuals will seek a formal education while others will rely on reading and personal investigation. Thought processes are directed towards organizing information into structures that are more understandable, practical and realistic. All forms of organized communication are emphasized, including writing and teaching. Occasionally a feeling of stupidity prompts the need to study a topic in greater depth.

For a limited few, normal pessimism can grow into depression. Saturn represents the perception of reality. For those who have built their lives around fantasy, this perception may be too stark or too structured. They may not want to see life as it really is, or they may not know ways to change it. The inability to seek alternative solutions to problems is the origin of much of the stress associated with Saturn-Mercury combinations. The need for restructuring the mental perspective is evident. Keep after problems until they are solved. Producing tangible results is very rewarding.

MERCURY-JUPITER ASPECTS

While Saturn-Mercury aspects are associated with pessimism, Jupiter, on the other hand, can imply optimism and confidence. You may look forward to the future with great enthusiasm. But what you believe to be possible may differ from your actual experience. Optimism can lead to miscalculations and poor decisions if you overestimate your chances for success or underestimate the amount of time you will require to complete a task. Overscheduling is directly related to this inability to foresee possible future difficulties. You might assume that tasks are quite simple, when in fact they are very difficult. You may not be able to honor deadlines and promises you have made if they are unrealistic. Stress results from these miscalculations.

For some, stress may also result from a lack of congruity between more philosophical beliefs (ruled by Jupiter) and daily experience (ruled by Mercury). You might find it difficult to believe what you see and hear. Situations you are involved in can directly conflict with long-held philosophical, ethical and religious beliefs. You may not now practice what you once preached. Very narrow minds will steadfastly hold on to basic beliefs which are contradicted by personal experience. These individuals will be unable to make the philosophical adjustments necessary to accommodate new information. Do not allow intolerance and hypocrisy to limit your possibilities. This is a good time to expand your mind. Return to school, join a discussion group or study on your own.

MERCURY-MARS ASPECTS

Mars-Mercury aspects suggest an energetic thought process. This can be a time of great mental energy and an active search for knowledge. Your mind should be quick and alert, though not necessarily highly retentive. Learning can be very exciting and self-perpetuating even if you study alone. What is great for the learning process may not be so advantageous when making decisions. The speed normally associated with Mars may indicate that you are impulsive and quick to jump to conclusions. You may not take the time nec-

very for careful consideration or thorough research. Instead, you may choose to handle situations with gut-level reactions rather than considered responses.

Mars-Mercury combinations imply assertive ability, but at their worst, these aspects can suggest aggressiveness and great anger. Ongoing conflicts and daily confrontations may occur with very negative manifestations. Those with strong tempers may have trouble controlling their anger. In the heat of the moment, they will say things without thinking about the consequences.

If the chart interpretation indicates unconscious pressure along with strong Mars-Mercury aspects, old unconscious anger and resentment may fuel present conflicts. You may not be fighting for what you believe is right in this situation; instead you fight because you have been wronged in the past. Focusing on angry thoughts and acting out conflicts can be a terrible waste of a good mind. Mars-Mercury aspects show that thoughts can be put into action. Get motivated. Use this time to accomplish many things. Let your experiences teach you along the way.

Mercury's relationships to the other planets in the solar return chart indicates how information is gathered, assimilated and integrated into attitudes that persist for most of the year. How easily this is accomplished, and in what manner, is suggested by the aspects. What you are actually thinking about is indicated by Mercury's house placement. The important thing to remember about Mercury's placement is that it is indicative of a mental exercise only. Alone in a house, Mercury can show mental preoccupation without psychological pain or physical consequences. It suggests the ability to make decisions only, and may not be a clear indication of action in any particular area.

MERCURY-VENUS ASPECTS

When Mercury is conjunct Venus in the solar return chart, you are more apt to have confidence in your decisions and intellectual capabilities. This is a wonderful time to be in school or to take a course since learning is likely to be an enjoyable experience. You may also make great progress in what most people consider to be a difficult topic or a subject you have had particular trouble with in the past. Facts can be easier to comprehend during this time and the information you receive should be applicable to your present needs or skills.

Your verbal tactics must include good negotiation skills. Develop a soft style for expressing your needs and getting what you want.

Learn to compromise and generate win-win solutions to problems. By doing so, you will be able to retain an inner peacefulness and a relaxed state of mind.

This is a good time to cash in on money-making ideas or save money by following a financial plan. If Venus and/or Mercury are associated with any of the work houses (2nd, 6th, or 10th), your ideas might generate funds for you personally, or for your company. Your ideas might also save money by cutting expenses or streamlining operations. Resources are not likely to be spent freely, but according to a plan with a goal in mind.

Mercury in the Solar Return Houses

MERCURY IN THE 1ST HOUSE

Mercury in the 1st house can show that your mind is very focused on self-interest. You will concentrate on your own needs, develop your own ideas and opinions, and/or make your own personal decisions. You will tend to be intellectually independent. You will rely more and more on your own thought processes and make decisions independently of feedback or consultation with others. Subjective interpretation can be very strong. You may only see one side of an issue, especially if there are no indications of objectivity in the solar return chart (e.g., 7th house planets and oppositions). Understanding other viewpoints may be difficult if you cannot relate to ideas other than your own. You value your own thoughts and opinions so highly that you may believe what you want regardless of what others tell you. This may not be a good time to truly understand others. You might have blind spots which make you unyielding and uninformed. You may stubbornly insist that you are right without reviewing the information. This can be a time of great conviction or great stupidity.

Intellectual development is possible during this time, but generally the individual prefers to be self-taught rather than returning to school. More than likely, the emphasis is on testing and using already acquired intellectual abilitties rather than focusing on further development. This is a time to put into action what you already know. Mercury in the 1st can be very good for reading, writing, studying or any task that requires single-mindedness to complete. Even so, sometimes there are two major tasks being worked on.

Your mind is very active and very quick, perhaps even impulsive. You may choose to hammer out your ideas during conversations. Be aware that you will tend to be very opinionated and perhaps even dogmatic during the discussions. Because you are in the process of developing your ideas, your thoughts will change over the year. You may have to eat some of your dogmatic decrees, so it would be wise to talk softly.

If you are not learning, reading, writing, studying, or actively thinking in a productive manner, Mercury in the 1st can symbolize a negative use of mental energy. Misdirected thought processes can succumb to nervousness, anxiety and depression. Inconsistencies between what you know and what you are told may be at the root of your anxiety. It would be better for you to get away from a preoccupation with your problems, and to focus instead on positive mental alternatives and solutions to your situation. This can be a highly productive year, one in which your personal preferences gain strength and significance.

MERCURY IN THE 2ND HOUSE

Mercury in the 2nd house is associated with financial planning and monetary decisions. Planning usually involves immediate financial needs rather than long-range retirement plans, which tend to be more closely related to placements in the 4th house of the solar return chart. More specific to Mercury in the 2nd house are conscious decisions about expenditures and income, and discussions about establishing a household budget. You need to discover how your money is being spent and may keep detailed lists of expenditures. If you have been an impulsive buyer in the past, you may decide that now is the time to carefully assess the necessity of each future purchase.

You may also want to make decisions about the amount of money you are earning. It may be necessary for you to get a job if you are not already employed, or earn more through overtime or part-time work. For those who are already working, this is a time to reassess the amount of money you are earning in relation to the quality of work you put out. The 2nd house rules one's sense of self-worth. Mercury in the 2nd can indicate that you decide that you deserve more money (rarely less), not because you need more or less to live on, but because your salary should be commensurate with your value as a worker. You are apt to be very vocal about renegotiating your salary. This is a good time to get your facts straight and graphically illustrate your increased productivity. Keep a log of money-saving ideas you have generated for your company and the amount of money you were able to save. Be certain of your facts and don't overestimate your productivity or worth. If you are self-employed, run a cost analysis of your office expenses and raise prices accordingly.

Self-worth issues are also common for those who are not employed or who only do volunteer work. You may feel that the services you offer to others are taken for granted. You need to hear how much you are appreciated. If you reassess what you do for others in regard to how well you are treated by these same people, you may decide that you should expect better treatment. Encourage others to appreciate what you do for them. Also learn to appreciate yourself. If time spent helping others drains away time you should spend on your own needs, it is essential that you prioritize tasks so that all the essential ones are completed first. Do not shortchange yourself. Your needs may be as important as the needs of others. This is a good time to make changes because you deserve more consideration and better treatment. Decide to limit your contact with people who do not respect your needs, to avoid situations that are not good for you.

MERCURY IN THE 3RD HOUSE

Mercury in the 3rd house suggests a strong inquisitive mind with a thirst for knowledge. You could spend a lot of time reading, writing, studying, or communicating. You want an influx of new ideas. Information that you gather during this year may relate to a project or particular field of interest. But it is also possible that your interests are scattered and the information is superficial. If you are attending school, learning may tend to be tedious and involve the memorization of numerous details. Interpreting Mercury's aspects and assessing the mental conditioning reflected in the chart can give you a better idea about what type of information is important and its purpose during the coming year.

All types of information can be important. Significance is not limited to educational material. This can be a time for major disclosures and realizations. It is possible that the information you receive at this time or have received in the past is false. You may be misinformed or even lied to. Mercury in the 3rd does not guarantee that the information you gather will be correct. There may be inconsistencies between what you are told and what you intuitively feel. It will be your task to assess what is truth and what is fiction.

Mercury in the 3rd house can suggest that the rational thinking processes may be stronger than feelings. This is most likely to be so if the Moon is not particularly strong by house, sign or aspect, and the chart is not watery. If the greater emphasis is on Mercury, decisions will be based on rational considerations rather than emotional needs. But if the Moon is very prominent in the chart, rational thoughts may be overwhelmed by emotional considerations. When the interpretation for the Moon (and/or Pluto) appears to conflict with the interpretation for Mercury, emotions and rational thoughts may seem to contradict one another. You may be torn between what you know or hear and what you feel in your heart or sense intuitively. If you are very stressed, unconscious needs may surface. Negative feelings, obsessions, compulsions, phobias, and extreme anger may defy rational control.

Your mental attitude during the year is very important. You need to think clearly in order to function at an optimal level. Aspects to Mercury in the 3rd house can suggest influences that either hinder or promote logical thinking. Among the more detrimental influences are abusive substances such as alcohol and drugs. Mental illnesses (especially depression and anxiety) are also negative influences which can affect one's ability to think clearly. These very negative manifestations correspond more closely to an individual's negative behavioral pattern than to any one specific astrological pattern in the solar return chart. Although stress may be suggested by the solar return chart, disease is not. Those who are intimately and enthusiastically involved in life experience few problems. For them, great excitement rather than stress fuels their thinking processes.

MERCURY IN THE 4TH HOUSE

Mercury in the 4th house suggests that you should make long-range decisions regarding your future financial security and stability. You need to provide for your own retirement through a savings account or retirement plan. Reassess your present financial arrangements to see if they are adequate for your future needs. Mercury by itself does not indicate difficult problems in these areas, only a mental focus on a secure future.

You are also likely to make decisions about your home or present living conditions. Will you fix up your present home or will you move to another? Is your neighborhood run-down? Does your present home meet your immediate needs and will it also meet your future needs? This is a good time to make long-range plans and assessments which will affect your living conditions this year and in coming years as well. Decisions will be based on a rational reassessment of your present situation and a need for domestic contentment. If you do not presently own a home, you may conceive a plan that will make home ownership possible. Regardless of where you live, you will want to analyze your use of space or redecorate. Any type of redecoration or renovation usually involves great planning and will not be done on the spur of the moment.

You will reflect on your past, your childhood and your relations with your parents. You may learn new information which will give you a different perspective on past events, feelings and relationships. If your parents are elderly, you may have to make decisions for them concerning their care and future security. Communication with family members may be important. You can coordinate or take part in a family project. If you are alienated from certain relatives, this may be a time when you wish to reopen lines of communication.

If you are a parent yourself, you should be concerned with your ability to parent effectively. This is a good time to gather information or join a discussion group that emphasizes the techniques and problems associated with raising children. Become aware of your assets and shortcomings as a mother or father. Realizations can lead to a search for new information and new ways to handle situations.

Mercury in the 4th house can also indicate an emphasis on the relationship between the conscious and unconscious mind. You may be actively investigating and analyzing your emotions. As the year progresses, it becomes easier to discuss those feelings with others. Obviously this is an excellent time for therapy, especially for those who find it difficult to think clearly and logically. Psychological complexes may intrude on rational thought processes. For some individuals, conscious/unconscious communication will include obsessive thoughts, irrational feelings and anxiety attacks. These conditions may be mild and involve a need to listen to suppressed information or emotions. Vocalizing your true feelings will help.

Mental difficulties associated with Mercury in the 4th house are generally confirmed by heavy emphasis on the Moon, Pluto and the 3rd/4th house in the solar return chart; however, these placements can also indicate a well-integrated conscious and unconscious perspective on life, especially if one is a writer, therapist or very awa

MERCURY IN THE 5TH HOUSE

A 5th house Mercury can indicate an increased interest in creative pursuits. This is a good time to express yourself artistically or creatively. Writing articles or books, composing music or poetry, painting or sculpting will channel your self-expression into artistic media. You are not necessarily any more creative this year than you were last year, but you have a stronger need for self-expression. The key here is to find some positive way to use your mental abilities in expressing yourself more fully. You do not have to be artistic to do this. Generally it is the expression itself that is more important than the means. There are many mundane media. You may plan out the landscaping around your house; you may create lesson plans for a course you are teaching; you may start a newsletter concerning something that you believe in passionately. Failure to find a good forum may mean that you risk becoming very opinionated and domineering in your personal relationships. Being pushy is a poor substitute for creative self-expression, and the desire to dominate others should be consciously avoided.

Your thoughts may center around romance and sexual encounters. If you are involved in a strong infatuation this year, you may find it difficult to concentrate on work and studies. Romantic preoccupation is possible. You may spend a lot of time analyzing the positive and negative qualities of this interaction. Relationships formed at this time are more likely to have good verbal rapport. Intellectual stimulation and communication may play central roles in your attraction to this person. However, this does not guarantee a good emotional relationship, only that the two of you will find it easy to talk and exchange ideas.

If you are a parent, this is a good time to focus on communication with your children. Situations involving your children will make you more aware of their mental health, educational needs and intellectual functioning. You may need to make decisions concerning their future. If they are older, you will have to let them make their own decisions; however, this does not mean that you cannot offer an opinion.

MERCURY IN THE 6TH HOUSE

Mercury in the 6th house can indicate a desire to evaluate your health and health practices at this time. Mercury by itself does not generally indicate serious health problems, but rather a realization that changes should be made if wellness is to continue. This is an excellent time to make decisions that will have a positive effect on your future health. You should become more aware of the value of exercise, adequate rest, good nutrition and eating habits. Educate yourself on these topics and begin to incorporate the information into your daily routine. Learn stress-reduction techniques and question your involvement in stressful activities. Your mind is not only instrumental in making decisions concerning your health, it is also directly related to physical health. Stressful situations can quickly lead from nervousness and anxiety to physical illness; therefore relaxation and the elimination of unnecessary stress is crucial. If your mind is not focused on positive learning, you can become very anxious about your health. Excessive worry can lead to hypochondria. Choose to work with health issues before they become health problems.

During the coming year, you may have to face certain facts about your job, work habits or working conditions. It is likely that your job will grow a bit more tedious and detail-oriented. You may have a lot more paperwork to handle. Your ability to pay attention to detail may be helpful if you are working in quality control, but do not let your push for the perfect product become stressful or obsessive. Tension on the job, without recourse, can make you overly critical of your position, co-workers or employees. Use your analytical ability for constructive criticism. If you are in a position of authority, you may be able to reorganize your office or working time so as to be more productive. The 6th house rules the day-to-day issues surrounding employment, rather than major career decisions. Analyze the daily operation of the office and explore methods for greater efficiency. If you are not in management, you can still make suggestions for improvement. Even those individuals who have no authority must make decisions concerning their jobs or work environments. Depending on your position within the company, you may not have a lot of freedom in these decisions and may only be offered "either-or" choices.

MERCURY IN THE 7TH HOUSE

A 7th house Mercury in the solar return chart can signal the importance of communication with others during the coming year. You may spend a lot of time discussing, explaining, negotiating and even arguing with others, especially if you have a major project or issue that needs to be handled. Matters under discussion will not

finalized easily. Projects may require much communication among all those involved before completion; disputes may require lengthy negotiations before agreement can be reached. Other planets in the 7th house or planets aspecting Mercury may give you some clue as to the nature of these discussions. Topics may relate to either personal or business matters.

On a personal level, communication within couples will be important. This is a good time to share thoughts and get in touch with one another. You may want to attend a marriage encounter. If you are having marital difficulties, see a marriage counselor and open up the lines of communication. Clarification and negotiation may be needed to improve or restore your relationship, especially if you are aware of contradictions and discrepancies.

Professionally, you may negotiate and communicate directly with others or hire the services of a professional to act as a go-between. Mercury in the 7th house can be indicative of another person with a great deal of knowledge or expertise. This is a good time to seek the advice and professional assistance of someone who is viewed as an authority in your particular field of interest or problem area. It may be in your best interest to let this expert handle certain business matters for you, but it is still important that you oversee proceedings and discuss decisions being made.

Both personal and professional relationships may be valued for the information and intellectual stimulation they provide. You will be attracted to people who are bright and full of ideas. The mental interchange fostered can lead to great creativity and insight through combined efforts. These results may far exceed the intellectual abilities of the individuals working alone. This is a good time to share ideas with others or to join a discussion group for the purpose of intellectual growth.

The danger associated with Mercury in the 7th house is a tendency to let others dominate your thoughts with their own opinions and comments. Do not personally accept truisms without assessing their validity or practicality. Since you are very open to advice from others, you may lack conviction in your own mental competence. You may grow dependent on the decision-making abilities of others. Mercury in the 7th house can indicate that you are lied to during the year. There is no guarantee that the information you receive is accurate or helpful. Therefore, it is essential that you evaluate information and advice as you receive it. This is an excellent time to gather information from

others, but it is still your responsibility to evaluate the informati
received and make your own decisions.

MERCURY IN THE 8TH HOUSE

The phrase "acute awareness" encapsulates the interpretation of Mercury in the 8th house of the solar return. Your mind is mo insightful and reflective during this year, and there will be tim when you will see more or know more than you were previously aware of. Knowledge of the unconscious mind grows quickly, lea ing you overwhelmed by the change. You are more aware of nuance motives and psychological conflicts in yourself and others. Subtle shifts in power and power plays will be more obvious to you now. Th interpretation of Mercury in the 8th house is very similar to th interpretation of the Sun or the Moon in this house, but there is an important difference. When Mercury is in the 8th, you are more ap to gain understanding of psychological issues as a result of trainin and education. You become intellectually aware of psychologica complexes and power struggles, but they generally do not disrup your life or upset you emotionally. When the Sun is in the 8th hous you are more likely to be disrupted by the turmoil of living with your own psychological issues or those of others. When the Moon is in th 8th house, you may be upset emotionally by complex unconscio forces and power struggles. Mercury in the 8th house is generally not upsetting, only insightful.

There are several ways to initiate and stimulate this psycholog cal insight. You may study psychology directly or you may becom involved in one of the occult sciences. You may see a counselor durin the year even though you might not feel seriously stressed. Mercu in the 8th house implies the ability to grasp psychological informatio intellectually. Generally, by itself, Mercury is not indicative of ner vousness or anxiety. If you are just entering therapy at this time, yo may be doing so to gain self-knowledge rather than because of incapacitation. If you have been in therapy for a while now, this may be the year when therapy pays off and suddenly you begin to see th behavioral patterns that have been affecting your life negatively.

Psychologically you may be more withdrawn and cautious dur ing the year, preferring to talk to one person in depth rather tha several friends superficially. Small talk will not interest you. Informa tion that is factual and blatant will not be as exciting as what is less obvious, sexual or mysterious. You see and understand a lot mo

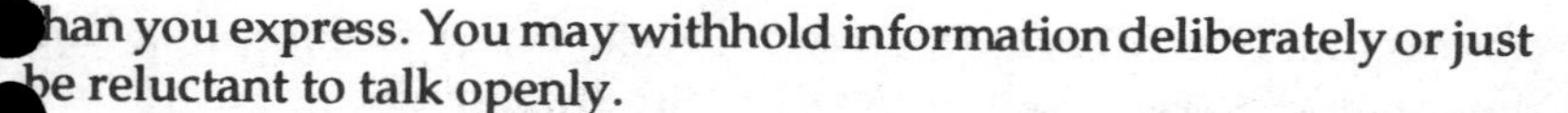

than you express. You may withhold information deliberately or just be reluctant to talk openly.

This is a good time for financial planning and consolidation loans. If your debts have risen recently, you might want to establish a payment plan for reducing those bills. If you share income and expenses with another person, review the division of debts and plan for future expenditures. Reread your will and investigate estate and tax plans.

MERCURY IN THE 9TH HOUSE

A 9th house solar return Mercury suggests an intense learning situation. You may return to school or continue your education in one form or another during the coming year. Education may involve a formal teacher-classroom setting; or you may study independently or take a correspondence course. Topics are diverse and not necessarily limited to a college curriculum. It is the keen interest and the concentrated study that usually characterize Mercury in the 9th house, not the topic of interest or classroom attendance. Even those who study on their own tend to be very interested or even obsessed with a specific topic.

It is also possible that you are the teacher rather than the student. If so, you are probably just starting out in the profession or need to rework your lesson plans. Your interest in the subject matter is usually high and you make a conscious effort to present the information enthusiastically. Unenthusiastic teachers with this placement may need to go back to school and recapture the thrill of learning for themselves. They may also need to reassess their beliefs concerning the teaching profession.

Teachers will not be the only ones reassessing their beliefs in regards to their reality. Everyone with a 9th house Mercury will be involved in this reevaluation process and all types of beliefs will be evaluated. The 9th house rules mundane beliefs as well as philosophical. Misconceptions, prejudices and unrealistic expectations fall under the rulership of the 9th. If you are involved in a difficult situation and you have any planets in the 9th house of your solar return chart, you may have misconceptions that are contributing to your problems or complicating the issues. Mercury in the 9th indicates that the assimilation of new information may help you to deal with these problems or issues. What you learn from your educational pursuits may directly apply to your present situation. This is the time

to use your education to reassess your beliefs and philosophy of lif
Situations you are involved in will naturally stimulate philosophica
adjustments, but you can foster this reevaluation process by actively
seeking information directly related to present situations or issues

If you have written a book or plan to write a book, this may be good time for you to approach a publisher or contact a literary agent. Communication with those in the publishing business can help yo with your project. Generally, though, publishing itself is indicated b more than the placement of Mercury in the 9th house. Usually the Sun in the 9th or a 9th house stellium is more indicative of publishing. you view your book as a career achievement, placements in the 10t may suggest publication.

This is also the case with traveling. Mercury in the 9th by itse does not generally imply travel. A 9th house stellium, Venus o Jupiter are more likely to do so, especially if you are looking for indications of foreign travel. Moon in the 9th may indicate livin overseas for a period of time or visiting foreign relatives.

MERCURY IN THE 10TH HOUSE

Mercury in the 10th house is associated with decisions that wi channel the direction of your life over the next few years. During th year, you will be asked to make choices that will affect your immed ate and distant future. Decisions may be career-related and directl involve your present job or professional goals. You might decide to attend school or take a refresher course. On-the-job training an work-related travel are also a possibility. On the other hand, deci sions might pertain to major life transitions without an emphasis on career. Important decisions such as marriage, leaving home or mo ing to another country can be indicated by Mercury in the 10th hous People who decide to change their lifestyles totally can have this placement.

Professionally, this can be a good time to communicate with yo superiors. Networking can further your advancement. Make busi- ness connections now and plan for future career developments. Yo may have ideas that would improve business or streamline offi procedure. These creative ideas can play a crucial role in career developments over the next year. If you are already in a position authority or willing to accept more responsibility, you might be ask to play a greater role in the decision-making process at your place of employment. If you are not in a position to change office routin where necessary, work may become more tedious. Tasks may be predominantly detail-oriented or repetitive and involve a great deal of paperwork.

MERCURY IN THE 11TH HOUSE

The 10th house rules laws and standards established by society to govern groups of people. Because they were prepared for the masses, they may not fit all individual situations or meet individual needs. The 11th house rules laws and standards established by the individual after reevaluating society's restrictions and regulations in light of personal experience. With this Mercury placement, you will probably question social restrictions in an attempt to personalize limitations and understand their usefulness in your present situation. It is possible that your needs as an individual conflict with the society's expectations. For example, if you are involved in a very difficult marriage and you are attracted to someone new, you might toy with the idea of having an affair. Pressure and needs relevant to your personal situation may contradict social restrictions. Because Mercury usually relates to a mental exercise, it is not necessary that you actually transgress social norms or experience compromising situations. Your reassessment may consist of a continuing intellectual or theoretical debate running through your mind for most of the year. But it is essential that you question the rules by which you live your daily life, and that you develop a new personal code of ethics.

The 11th house also emphasizes group needs versus individual needs. Therefore it is common for the individual to be involved with a group or social situation which focuses on balancing these two energies. You may be involved in a group because you are more likely to attain goals through a combined effort than working independently. You may need to share thoughts and experiences with others and establish new goals for the future. The group you are involved in may be primarily intellectual or social, but it is also possible that the group is working towards a common goal. Communication among group members or the discussion of pressing issues may be an important part of the group efforts. Self-help groups may pertain to Mercury in the 11th house, although these kinds of groups are more closely associated with the Moon or Mars in the 11th.

Some organized groups tend to support group needs and goals over the needs of the individual. The individual may feel compromised in this situation. For example, a new collective business group hired a full-time advertising person to promote their business. They needed to hold down expenses since finances were limited. The person creating and placing the ads believed in the group efforts, but he needed to live off the salary received. In this case, the group goal of holding down expenses was not consistent with the individual's need to earn a good salary and provide a valuable service. The balance of individual needs versus group needs is usually emphasized in some way when Mercury is in the 11th house.

A question of personal freedom underlies the issues of Mercury in the 11th house. Whether you are dealing with friends, groups or society at large, the amount of individual freedom you enjoy is controlled to a certain degree by the relationships you have. You must accept certain restrictions and considerations so that your behavior does not impinge on the rights and freedoms of others. Therefore issues related to monogamy in a relationship or loyalty among friends may be important.

Since this is a decisive year for goals, the process of questioning established norms is particularly important. This is a time when you should be open to the new ideas and directions so necessary to the formulation of future goals. The questioning process opens your mind to original ways of thinking and frees you from unnecessary restrictions. The reevaluation process also helps you distinguish practical goals from unrealistic dreams. Mercury in the 11th house can suggest that you are able to justify theoretically your need for freedom from pointless restraint in order to pursue new goals for the future.

MERCURY IN THE 12TH HOUSE

Mercury in the 12th house indicates that you are probably not outspoken this year. You keep your opinions, thoughts and feelings to yourself. You also keep secrets. You may be spending more time alone, lost in your own thoughts. This is a time for the development of new ideas and the understanding and organization of what you have already learned. Consequently, you may not be sure of exactly what you want to say or how to say it. Unfortunately, you may not be saying some of the things that need to be said. Mercury in the 12th house is associated with keeping quiet to preserve peace. You may find that when you do express yourself, it causes an argument. So it seems easier to compromise before the argument starts and just bite your tongue to begin with. You may not verbalize your true thoughts even when you should. If you are caught in a lie, even one which was perpetrated by others, you may not correct the situation by telling the truth as you know it.

This is a good time to be reflective and introspective rather than just withholding. This can be a very religious or spiritual placement for Mercury. Meditation, quiet reflection and spiritual studies are important and may give you a greater understanding of the Universe and your place within the Universal plan. But the concept of the "big picture" is applicable to all information and not just that which is spiritual in nature. You have accumulated much information in recent years which now needs to be organized into accessible and practical knowledge. Let's use an analogy as an example. If you are an astrology student, you spend many years learning astrological techniques. These techniques provide you with different bits of information. At some point you will need to learn to organize these pieces of information into a chart interpretation. This is the difference between "cook-booking" astrological facts and understanding the individual's personality by interpreting the whole chart. There is a leap of consciousness described here that transforms separate facts into cohesive knowledge. Mercury in the 12th can denote this leap of consciousness from facts to knowledge. It represents the organization and assimilation of data necessary for true learning. Private study and personal research can help you with this process, but inner thoughts should be your focus of attention. This can be a good position for the writer, artist or anyone beginning a long-term project which will involve much contemplation. Do not waste all this inner energy on daydreaming and fantasizing; use this time to deepen your understanding of what you have learned.

The review and organization of information is not limited to factual data only. Unconscious impressions, feelings and intuitive insights will also be important. The 12th house seems to relate to the integration of all forms of existing information, whether on the conscious or unconscious level. If there is a conflict between what you are thinking rationally and what you are feeling emotionally, the integration process may be more difficult and you will probably experience some anxiety. This lack of continuity may be due to your avoidance of certain issues and feelings in the past. But it can also be caused by inaccurate information you are receiving at the present time. A very negative manifestation is that someone is lying to you or purposely misleading you. Your task during the year to resolve any conflicts between feelings and conscious thoughts and become an integrated whole.

If your mind does not have a strong inner or outer focus this year, you could experience mental problems in the form of phobias, compulsions, obsessions, jealousy and anxiety. Free-floating anxiety (generalized fear) is the more common difficulty. Those who were counseling candidates before Mercury was placed in the solar return 12th house are not likely to improve on their own this year. If you are already experiencing problems, this is a time for therapy. The emphasis should be on integrating unconscious and conscious energies. Wisely channeled, you can use this time to intelligently evaluate unconscious insights in the light of previously gathered information. Together the two sources, the conscious and the unconscious levels, can provide you with a wealth of information organized into a body of knowledge.

CHAPTER FIVE

VENUS IN THE SOLAR RETURN CHART

Introduction

Venus has a repetitive cycle of eight different placements recurring in eight consecutive solar return charts. For the first seven years of your life (your natal chart plus seven solar returns up to and including your seventh birthday), you will have new solar return Venus placements each year. When you are eight years old, your solar return Venus returns to its natal position and the cycle begins to repeat itself. Venus continues to return to its natal position on those birthdays that are a multiple of eight (i.e., 16th, 24th, 32nd, 40th, 48th, 56th, 64th and 72nd birthdays, etc.). The Venus placements are repetitive plus or minus a few degrees. However, there is an exception to the rule. If your solar return Venus placement is retrograde or even close to being stationary, the degrees of that Venus placement will show greater variation from cycle to cycle. In fact, what was originally a direct Venus may become a retrograde solar return planet after many, many cycles and gradual placement drift.

The Venus cycle has important implications with regard to solar return interpretation. Each of the eight placements that appear in your solar return has a specific and recurring relationship to your natal chart. Analyzing the aspects between Venus and the natal chart may help you to understand your pattern of relating during the coming year. Carrying this understanding one step further, you may be able to correlate events and relationships occurring during any one year with those happening eight and sixteen years previously. Comprehension evolves from the interpretation of the single placement into an understanding of the eight-year connection. The cycling Venus placements seem to imply that relationships are evolutionary within our lifetime. As we mature, we continue to deal with the same relationship issues and feelings (though situations and experiences change somewhat), but with each cycle, we move to a higher and more evolved level of relating. Understanding the cyclic nature of Venus and its meaning within the context of our solar return chart should foster further growth in this area of development.

If you have a Venus placement that drifts because of proximity to the planet's stationary point, changes in the recycled Venus can be significant and this placement will continue to drift cycle after cycle. The years in which Venus placements change signs or directions (from direct to retrograde or vice versa) should be very important with regard to your love life or financial situation. These years should be investigated since they are a rare occurrence within any lifetime and do not happen to everyone.

Interpretation

Venus is the key to understanding your relationships and financial situation for the solar return year. Many times these two areas go hand in hand; financial security tends to be coupled with secure relationships while financial difficulties are more likely to occur when close relationships are strained. Although this correlation is common, it is not the rule, and interpretations for finances and relationships should always be viewed separately. When assessing Venus in the chart, evaluate Venus twice, first with respect to the other relationship factors in the solar return and then with respect to the other monetary indicators. In this way, you can develop a clear picture of each area of concern.

Venus also indicates a need for comfort and it is informative to note what has preceded Venus in its present house position. If, during this year, Venus resides in a house which last year represented an area of conflict, Venus would imply an improvement in the condition associated with that house and may indicate a healing process. An example of this might be helpful. During the first year of writing this book, I had Saturn, Mars, Uranus and Neptune in the 5th house of creativity. Writing was a very labored process requiring much time, effort, and editing. Chapters took months to complete. But I continued to write and sought helpful advice from editors, authors and friends. The following year, Venus and Mercury were conjunct in the solar return 5th house. By this time, the whole writing process grew much easier, but only because of the hard work during the previous year. Benefits commonly follow after much hard work. A Venus placement in the solar return chart following a more difficult placement can indicate great reward. Your awareness of this possibility may encourage you to work harder in more problematic areas since you can see the rewards in the following year.

Relationships

Venus, by house, shows what relationships will be important during the coming year. For example, Venus in the 3rd house might indicate that relationships with neighbors or community involvement will be emphasized. With Venus in the 6th house, pleasant office conditions and good co-worker relationships will be important. But the interpretation of relationships as they appear in the solar return chart involves more than an understanding of Venus' house placement. Look at the 5th and the 7th houses, especially if you are interested in a love relationship. The more planets in these houses, particularly the 7th house, the greater the need to relate on a one-to-one basis. Generally, the 5th house shows sexual affairs, while the 7th house indicates a greater commitment (though not necessarily marriage). Squares between the 5th and 7th houses frequently show an affair in which partners disagree about getting married, or perhaps one partner is already committed. Clandestine affairs or secret relationships are more likely to occur when Venus and/or the Moon appear with Pluto in the 12th house. Other factors to consider in your assessment and interpretation of relationships include: aspects made to Venus, and which houses they come from; orientation of the solar return planets to the eastern or western hemisphere of the chart (implying independent or cooperative needs); sign and position of the Moon; any placements in the 4th house (indicating emotional needs); and the house location of the sign Libra and any Libran planets. All of these factors need to be considered along with Venus in your interpretation of relationships as seen in the solar return chart.

Finances

Venus also relates to money and finances and can be used to evaluate these circumstances for the coming solar return year. The house placement for Venus may indicate how you are most likely to generate income. This is sometimes true, but not always. Venus in the 9th implies a teaching salary, but if you don't teach, the emphasis will be on your beliefs concerning relationships. Don't stretch your economic interpretation of Venus' house placement. If the house position applies, the situation will be obvious. If the house position does not seem to apply to finances in particular, Venus will be more consistent with relationship situations during the year.

For finances, it is perhaps more important to look at the planetary placements in the 2nd, 6th, 10th and 11th houses, especially if you are interested in the amount of money you will be able to earn by employment. The 2nd house is the primary money house, but it also gives indications of spending practices. Any planets in the 2nd can reflect your salary and/or your spending practices (regardless of how much money you earn); consequently it is always possible to spend more than you make or make more than you spend.

Pluto or Saturn in the 2nd seems to show the greatest financial change. Pluto is sometimes associated with unemployment and the resulting loss of earning power. It can be seen in the charts of those people who voluntarily stop working or take a pay cut. Spending practices change either to accommodate the loss of income or to create a savings account. You can cut out unnecessary expenses and trim your budget during this time. Tight budgets tend to be common whenever Pluto or Saturn is in the 2nd house.

People with Saturn in the 2nd are generally underpaid but they persist in their jobs because of other nonmaterialistic forms of gratification. Monetary settlements from wills, lawsuits or divorces will probably be delayed until late in the year, unless their value is greatly reduced or Saturn transits out of the solar return 2nd house. Those with Saturn in the 2nd house of the solar return tend to be savers. Many times, they are considering making a major purchase in the near future, so they budget their income and live on less now in order to have the money needed for an anticipated expenditure.

Neptune in the 2nd house usually indicates monetary uncertainty. The situation is one of financial insecurity rather than actual lack or need. You might be self-employed, working on a commission basis, or expecting to be promoted or laid-off during the year. The emphasis is on uncertainty and it lasts for most of the year. Changes are not likely to occur until the new solar return chart becomes relevant.

Uranus in the 2nd house shows an erratic income with equally erratic spending practices. Impulse buying may be a problem. There are salary highs and lows or large expenses. Again, you may be self-employed or working on a commission basis. The trick here is to conserve extra money for those rainy days. Financial upswings are as likely to occur as financial problems. Uranus is often associated with breaks in employment such as a leave without pay, e.g., maternity leave.

Mars in the 2nd house implies that the more you hustle, the more you earn. This is especially applicable to those who are self-employed or working on a commission basis. The more energy you put into working, the higher your salary.

Another indicator of possible financial limitation, whether self-imposed or not, is the interception. Interceptions in the 2nd and 8th houses and/or the interception of Taurus anywhere in the solar return chart can show a lack or funds or a tight budget. Double chart tecniques (with the solar return chart in the center wheel and the natal chart placed outside) that produce interceptions in the 2nd/8th houses, or of the 2nd/8th houses, in either the natal or the solar return chart, can also show limited finances.

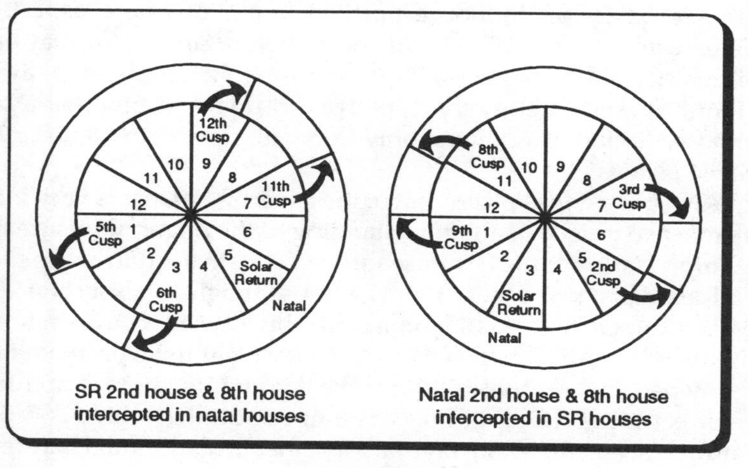

Figure 22

The 6th house is a work house and reflects employee relationships and daily working conditions and schedules. Planets in the 6th house that aspect Venus or 2nd house planets indicate that your income might be affected by illness, relationships with co-workers, the work environment, or profit-sharing. The unity and dedication of all employees pulling together may boost your financial compensation.

The 10th house is the career house and is closely associated with career moves, business reputations, and dealings with authority figures. Planets in this house can imply promotions and increased earnings, but since you can be very successful and even get promoted without getting paid extra, nothing is guaranteed. Planets in the 10th house that aspect Venus or planets in the 2nd house can show business practices that either increase your earning power or undermine your effectiveness. Your ability to please supervisory personnel or conform to company objectives may increase or decrease your chances for financial gain. For those who are self-employed, your reputation and ability to please the public will affect your income. Any type of career move is also likely to cause financial repercussions.

The 11th house is the house of promotions and money earned from the career (the 2nd from the 10th house). A planet in this house can increase or decrease the likelihood of a promotion or a pay raise. Goals are also associated with this house, and long-term goals may or may not support your present financial status.

Regardless of the house position, numerous hard aspects to Venus can imply complicated financial circumstances. You may have difficulties getting money or large expenses might drain funds away. Tight budgets, whether they involve a real lack of funds or a self-imposed limitation are commonly seen with squares to Venus or 2nd house planets.

Often the house placement of the aspecting planet is very informative and crucial to an understanding of the financial picture. For example, Saturn in the 1st house square Venus in the 10th house could indicate that a personality trait (1st house) might block a chance for greater career success (10th house). Ambivalence is common in this situation. Negatively, the individual may want to be promoted but heavy personal responsibilities make it very difficult to concentrate on career success. Despite mixed feelings associated with this Venus-Saturn placement, it is always possible that the individual may learn to develop the discipline and organization necessary to support career ambition.

Another example: Saturn in the 2nd square to Venus in the 11th house can indicate a pay cut or poor salary. The individual's low sense of self-worth may allow him or her to accept things as they exist. This person needs to realize the value of his or her talents and services. The same solar return placements can also show a native who budgets income very carefully so that the accumulated savings can then be used to attain a long-term, expensive goal. In each of these cases, the aspecting planet's placement directly relates to the interpretation of the individual's financial situation.

The type of hard aspect affecting Venus can denote the origin of the financial tension. Oppositions to Venus show two or more separate and distinct monetary goals. You may have more than one way to spend your money and not enough money to go around. Many times you will struggle with another person over how money should be spent, divided, or earned. You may argue with your spouse over bills, sue your ex-husband for child support or your doctor for malpractice, write your step-children out of your will, or declare bankruptcy and not pay your debts. These are extreme examples of monetary conflicts but are sometimes seen with oppositions to Venus and a heavy involvement of the 2nd-8th house axis. To work constructively with this aspect, negotiate financial decisions that are satisfactory to all the individuals involved and meet your responsibilities despite your limited means.

Squares imply tensions which affect your earning power. You may be unable or unwilling to match spending practices to funds available. Conjunctions indicate limitations that are self-imposed or the result of other choices you have made. For example, a person with a Venus-Saturn conjunction in her solar return charts for three different years had three separate problems associated with her ability to earn money. During the first year, she could not work because she was fulfilling full-time school internship requirements towards her college degree. During the second year, she moved to a foreign country where she could not work because of the language barrier. When she finally returned home, she started to work, but at a low salary.

It is important to remember that it is the preponderance of information that creates the interpretation and not any one single indicator alone. For further information, please interpret the whole solar return chart and relate it to your present financial situation.

Planets Aspecting Venus

Aspects to Venus reflect your ability and desire to exchange love and money with others and under what circumstances. This is a simplistic statement that becomes more complex when we look at the individual planets that aspect Venus.

VENUS-PLUTO ASPECTS

Pluto-Venus contacts show intense emotional involvements and complicated financial situations. New relationships can be especially compelling, and like the moth to the flame, you are drawn to particular individuals without understanding why. Unconscious forces play a major role in this type of relationship. Eventually you may discover that you have met the perfect mate, but in the beginning, the perceived loss of control and diminished rationality is disconcerting. Regardless of what you think or plan, you end up reacting to situations in a spontaneous and revealing manner. Your usual psychological defenses don't seem to work, while all your unconscious complexes are laid bare for the world to see; consequently, a persistent sense of vulnerability becomes coupled with your growing need for intimacy.

In both new and old relationships, Pluto-Venus aspects can indicate considerable emotional growth but not without a good understanding of the unconscious urges and psychological games that are impediments to intimacy. Awareness of these intimacy barriers is crucial to the Plutonian process, while learning to deal effectively with these influences affects the success of the relationship. Obviously, this is a good time for counseling of any kind.

Situational barriers include involvement with married, gay or bisexual partners, those living in a different locality, or those who realistically cannot be fully present or involved for one reason or another. These factors represent preexisting impediments to intimacy that were probably known or suspected before the relationship even began.

Psychological impediments to intimacy are those unconscious complexes which distort reality and destroy trust. These include, but are not limited to, possessiveness, obsessive thinking, sexual pre-

ances, compulsive behaviors and controlling attitudes. Potential or existing partners can exhibit a new or troubling flaw (which may or may not be serious). For example, a very extravagant young lady was totally in love with her boyfriend but troubled by his lack of wealth. She wanted him to meet her material expectations by earning more money; consequently, she sought to motivate (control and manipulate) his behavior.

Power can be an issue in both intimate and casual relationships. Individuals who learn to trust, compromise, and share power see their relationships deepen and strengthen. Accommodating the emotional needs of others breeds increased understanding. Those who are unable to reach compromises become locked in power struggles, and feel powerless to control their own destiny. They view their fates as dependent on the whims of others and resort to controlling behaviors as the only solution. Within this struggle, manipulative techniques are the main weapons; sex and money become the main issues.

Financially, Pluto-Venus aspects indicate strong financial changes or complex monetary arrangements. Salary changes are common and may be the result of a career move, relocation, leave without pay, cutback in hours, or retirement. Ambition can cause your salary to rise dramatically, especially if your earnings are based on commissions or profit-sharing. Those who are self-employed experience fluctuations in earnings. Relationship problems may directly affect your ability to concentrate and work, or your financial status may depend on the resources of others. Struggles over money are not uncommon. A college-bound student sought to manipulate her grandfather into paying for her education, while a married couple traded money for sex.

The emphasis here is on personal power in relationships and power over your own financial situation. By gaining insight into your behavior and the behavior of others, you can master both areas of concern.

VENUS-NEPTUNE ASPECTS

If you are already involved in a stable emotional relationship, Neptune-Venus aspects may be more relevant to financial concerns than romantic interests. Financial circumstances will exhibit a degree of uncertainty, but only rarely is a person unemployed for part of the year and not sure where the next dollar is coming from. A break from work is more likely to be a leave of absence without pay, but even this situation is uncommon unless you are female and pregnant (maternity leave). With Neptune-Venus aspects, it is more common to experience other reasons for financial uncertainty. You can be self-employed or have an income based on commissions, incentives or profit-sharing; your hours may fluctuate or you might work on an "on-call" basis. Under these conditions, you will be unsure of the amount of your pay until the check arrives.

Changing circumstances also affect the amount of money available (e.g., relocation, divorce, illness, large purchases or major sales) and it is impossible to predict the effect these changes will have, especially if several variables are involved. For example, if you must sell your home and relocate to a different state, you may not be sure how much you can sell your present home for, how much your new home is likely to cost, how much you will be able to earn at your new job and how expensive it will be to live in the area. You must be careful with your funds until the figures are worked out. If you are expecting money from others in the form of a loan, gift or inheritance, you will not be sure how much you will receive or when you will receive it. If you are taking a financial gamble, you can't be sure that things will work out.

Regardless whether or not financial uncertainty is job-related, monetary uncertainty exists in some form, but it need not be a serious problem. This is a good time to foster a strong appreciation of the nonmaterial side of life. Since material, monetary, physical and external indicators are so uncertain, focus on the inner qualities.

Neptune aspecting Venus can also be important to understanding your relationships during the coming year. Significant emotional relationships can be very compassionate or very confusing. In strong relationships, the partners are more likely to be sensitive to each other's needs. Understanding and empathy increase and inner qualities are the focus of attention and appreciation. You can be more aware of the subtleties of your partner's unspoken preferences. Sometimes you neglect some of your own needs so those of your partner can be fulfilled.

The ability to be self-sacrificing is seen with this aspect, but then, so too is the ability to be martyred. External trappings are devalued and what you materially own and share is not as important as what you are willing to share emotionally. For the time being, feelings may be all you have.

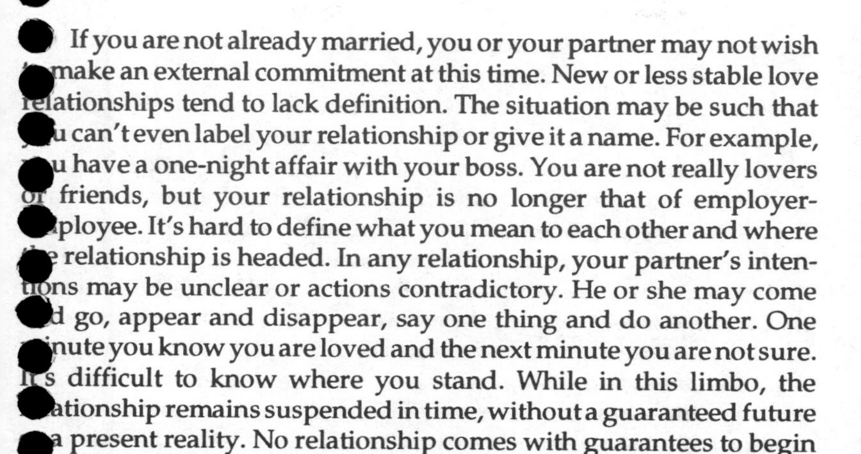

If you are not already married, you or your partner may not wish to make an external commitment at this time. New or less stable love relationships tend to lack definition. The situation may be such that you can't even label your relationship or give it a name. For example, you have a one-night affair with your boss. You are not really lovers or friends, but your relationship is no longer that of employer-employee. It's hard to define what you mean to each other and where the relationship is headed. In any relationship, your partner's intentions may be unclear or actions contradictory. He or she may come and go, appear and disappear, say one thing and do another. One minute you know you are loved and the next minute you are not sure. It's difficult to know where you stand. While in this limbo, the relationship remains suspended in time, without a guaranteed future or a present reality. No relationship comes with guarantees to begin with, but with Neptune-Venus contacts insecurity can be even more evident.

Because of the lack of clarity inherent in relationships, idealization and distortion occur, making partners seem too good to be true. Aggrandizement of this sort can only lead to disappointment and disillusionment. A persistent sense of confusion and insecurity can cause you to misinterpret what is actually occurring. This is most likely to happen when one or both partners are ambivalent or deceptive.

At best, relationships possess a sense of compassion and sensitivity. At worst, there are lies to contend with. Clandestine and secret relationships are common with Venus-Neptune aspects and 12th house placements. Lovers meet in private and keep their relationship hidden. Even if you have nothing major to hide, you may want to meet and be alone with each other.

VENUS-URANUS ASPECTS

Uranus-Venus aspects in the solar return chart usually denote changes in relationships. For some people, this can mean a sudden attraction and the excitement of a new love. Involvements may be very strong, very quick and may or may not have staying power. New relationships might eventually lead to marriage, but probably not during this solar return year. Loose associations are common. Friends become lovers and some love relationships seem more like friendships. Occasionally, relationships and attractions are less than conventional. Affairs between or with married individuals, homosexual contacts, and long-distance romances are possible. A general mood of disruption may permeate all partnerships, whether old or new. Difficult relationships may end or become on-again, off-again involvements, but for those in a healthy relationship, changes will occur within the relationship itself and do not indicate a break in ties or a loss of commitment to one another. One or the other partner has a desire for greater freedom and openness. For some people, the changes involve a loss of comfort and stability, but a gain in newness and excitement. For example, the birth of a child brings both disruption and excitement to the new parents. Gone is the relaxing comfort of a stable schedule; expectations and behavioral patterns which were once taken for granted may now be nonexistent. New patterns must emerge, but while the necessary adjustments are being made, the continuing disruption can cause difficulties and temporary distancing between the people involved. Any strong change can cause this disruption in established relationship patterns. Common examples are relocation, retirement, heavy travel, and changes in the number of people living with you.

Financial changes are also likely to occur with Uranus-Venus aspects. If you are involved in a new and exciting relationship, you may give up your present home and job to relocate to his or her area. Some individuals dispose of much of what they own before the move. For one reason or another, you may decide to stop working or you may go out and get a job. The amount of money available to you is likely to change or fluctuate. You can be working on a commission or incentive basis, or be involved in a profit-sharing plan at work. Self-employment is also possible with Uranus-Venus aspects, especially if you are newly self-employed and have not established a steady clientele.

VENUS-SATURN ASPECTS

Saturn aspecting Venus can have one of two different manifestations: the solidification of a relationship, or the inhibition of a relationship. Both processes involve the use of structure to define a relationship. Healthy relationships can strengthen and show a renewed sense of commitment to the future by both partners. The ability to accept more responsibility with or for each other is common. Marriage, raising children, and purchasing homes together are three common events that reflect this change. The relationship gains more stability because of a new definition of purpose and intent.

Partners who basically love each other and are committed to a relationship, but experiencing a period of change and adjustment, might do well to give each other a little space, understanding and support. The distance needed can foster individual expression and achievement. For example, if you or your spouse wants to go back to school for a higher degree, time must be set aside for classes and study.

Relationship inhibition occurs when the structure and definition of the relationship is used to limit rather than give a sense of purpose. You may not be allowed to give and receive love in a relaxed manner since love is always controlled, defined, withheld, rejected or not accepted as it exists. You yourself may refuse to make a further commitment, or it can be your partner who backs off. Statements like, "We're just friends," or "This is only an affair and I could never leave my wife," are two examples of the kind of structuring associated with Saturn-Venus contacts. Very poor relationships may be defined by separation and divorce. One person may accept total responsibility for a second individual. The "parent" partner acts as the person in charge of the "irresponsible child" and the relationship loses all sense of balance and equality.

Financially, you will probably be working with less cash during the year. Your budget will be tight, either because you have limited funds or you choose to conserve money and build savings. If you are normally very free with your money, this is a time when you are more careful with spending practices. Even the very rich will want to know how much they have and where it all goes.

Some may experience a cutback in funds. These are the people who quit their jobs, cut back on hours or salary, become self-employed, or experience a dramatic change in their lifestyle which would naturally affect funds available.

VENUS-JUPITER ASPECTS

Aspects from Jupiter to Venus have several possible interpretations. You can benefit, either directly or indirectly, through the influence or assistance of others. Relationships can be very rewarding and you reap what you have sown. If you have been helpful and understanding to others in the past, you can expect the same treatment now. Your lover, business partner, or spouse may be actively supportive of your goals, either financially or emotionally.

Indirect benefits can also come to you through your partnership. For example, your spouse accepts a job transfer to Colorado and you always wanted to live in ski country. You may buy a house or expensive ski equipment when you get there since large purchases and expenses are commonly seen with Jupiter-Venus solar return aspects. Possibly you spend more than you can afford.

What you believe to be true about most relationships can either be consistent with your experience, or in conflict with the reality of your situation. As would be expected of anyone, it is more common to notice the conflicts than the periods of consistency. Issues concerning morals, ethics and monogamy are important and you may discuss whether your relationship is an open or closed one. Either one of you can be involved in a second relationship which causes this issue to be raised. Defining the ethics of relating becomes a goal during the year.

VENUS-MARS ASPECTS

Mars-Venus aspects in a solar return chart imply that actions either support or conflict with a relationship's needs. It is important for partners to be supportive of each other's actions and to appreciate what is being done. For example, a couple combined their efforts and totally redecorated an investment property they owned. Each partner contributed his or her own talents and abilities. Their individual efforts complemented each other and got the job done.

Actions can be either conflicted or supportive. Relationships that are faltering to begin with become combative involvements in which the participants undermine independent or joint efforts and fight over money. Sexual attraction and satisfaction are important factors in both new and established relationships and contribute to or detract from their success or failure. Financial cuts are possible, but it is more likely that you must hustle for the money you do receive. The harder you work, the more you are likely to earn.

Venus in the Solar Return Houses

VENUS IN THE 1ST HOUSE

Venus in the 1st house is usually indicative of a gentle demeanor. You would not wish to hurt anyone's feelings over transgressions that have occurred in the past, nor for intrusions occurring in the present. It is common to be nonaggressive and even nonassertive with this placement, depending on Mars. Your gentle and nonassertive manner can leave you open to the demands of others, which may be overwhelming, especially if there are oppositions between planets in the 1st and 7th houses. If this is the case, your attention is divided between your needs and the needs of those you are involved with.

As much as you wish to assist others, part of your focus should be centered on meeting your own needs, and protecting your own interests from outside influences. You need free time and energy in order to "do your own thing." For this reason, Venus in the 1st house is associated with self-protection and balance. You may need to learn to protect your own interests from the intrusions, opinions and demands of others.

The most common form of self-protection associated with this placement is withdrawal from contact with others. You spend more time alone during the year for one reason or another. You probably like being alone and feel you are your own best company. You do not want to give up personal needs and interests just to be with others. Aloneness may be a simple matter of changing environment. Graduates leaving a college dormitory and living alone for the first time might have this placement, as well as individuals engaged or married but separated from their loved ones.

If there is a strong opposition from the 1st to the 7th house, aloneness may alternate with a strong emphasis on relationships, indicating a seesaw, all-or-nothing lifestyle during the year. But usually "aloneness" carries more meaning and greater importance than merely being by yourself. You may need time alone to take care of your own interests. For example, a new writer needed more time alone to write an intense novel. She enjoyed working with the fantasy she was creating. Her time alone was used to foster a creative and productive working environment. During the year she tended to avoid superficial and unnecessary social contacts.

Another individual was plagued by the demands of several neighbors. She was home during the day and constantly called upon for help. She did not mind helping others, but her nonassertive personality allowed the demands to become excessive and eventually she had little time to herself. Her neighbors took advantage of her, asking her to do things they would normally do for themselves. She began to leave her phone off the hook and pretend to be out when someone came to the door. She used aloneness as an escape.

The reasons you want to be alone should be indicated by the rest of the solar return chart. You may be running from a situation you find difficult to control or to a situation you wish to create. The interpretation is modified greatly by Mars, its house position and aspects. Generally a strong natal and/or solar return Mars indicates an individual who is able to defend his or her own interests successfully without escaping. This individual is more likely to be alone because aloneness has innate value during the year. Nonassertive individuals are more likely to see withdrawal as a form of self-protection.

Benefits, money and comforts are self-made during the year. Do not expect others to provide for you unless there are other indications in the solar return chart. This is generally not the time for obtaining from others, but rather a time to be good to yourself. You are able to advance through your own efforts, but to do this you must be more aware of your skills and more confident of your abilities. Your greatest opportunities are self-made, and your greatest achievements involve personal satisfaction.

VENUS IN THE 2ND HOUSE

One might think that Venus in the 2nd house of the solar return would be associated with increased money and materialism. Certainly you pay more attention to finances with this placement. But rather than increased money, Venus in the 2nd house is more likely to indicate easy money, the kind that requires no extra work on your part. The epitome of this interpretation is money that comes to you through welfare or unemployment checks. This is a possibility, although usually not the case. The most common manifestation is that money comes more easily because you cut back on your hours of work or refuse overtime. You may earn a little less, but the quality of your life seems more important than the extra income.

Venus in the 2nd normally indicates you are comfortable with the money you have or earn, so it is most likely that you anticipated this cutback in salary and either paid off your debts or made adjustments in your budget so the financial loss would not be noticeable.

Another common manifestation for this Venus placement is an increase in salary or funds available. As a rule, pay raises will not involve a great increase in time or effort on your part. If you work as a salesperson, you may land a huge account, thereby raising your commission without increasing your work load. If you are employed at a steady salary, you may be given a raise or bonus.

Another alternative is changing jobs and getting paid more for doing the same work. Your salary may remain the same, but funds become available through other means. Suppose you sell your home at a great profit and decide not to buy another home, or purchase one at a lower price. This leaves more money available without any extra work. Even if you do not get a raise, or do not work, small checks may come in the mail from unexpected sources (dividends, refunds, interest, etc.). Probably none of the checks will be large but you will find them helpful.

During the year, you should feel comfortable with the amount of money you have available to spend. You are unlikely to feel impoverished unless there are other indications in the chart. Aspects to Venus and other planets in the 2nd house may give you some idea as to what financial situations will evolve. Aspects between the 10th and the 2nd or the 6th and the 2nd generally imply that money is job-related. If Venus is heavily aspected from the 8th house, debts and shared resources will be an issue. An emphasis on the 11th house with Venus in the 2nd would indicate the desire to work with long-term financial goals (e.g., saving for a college education). Whenever Venus is heavily aspected in the solar return, financial situations may be changing or complicated by other interests.

You are in the process of reassessing your value system. Priorities may be changed or even reversed. What may have been important before is no longer crucial; what you once took for granted is now cherished. Usually the shift is toward an appreciation of inner beauty and the quality of life, but some people experience Venus in the 2nd house as a surge in materialism, especially when Venus is in an earth sign. You may not want to do extra work to increase your cash flow, so you will try to think of less taxing ways to get more money.

You might also experience a moral or ethical conflict at this time, especially if Venus is heavily emphasized. Venus in the 2nd shows that moral issues will be reassessed, especially the issue of monogamy or loyalty to one person. Contrary to what one might think, traditional concepts are not always stressed. The interpretation of Venus in this house seems to be most closely related to proclivities shown in the natal chart and the rest of the solar return. Ethical issues focus on monetary practices such as accurate expense account reporting, appropriate fees for services rendered, and the disposition of funds. Several aspects to Venus in the 2nd can indicate mixed feelings with regards to these moral or ethical issues.

VENUS IN THE 3RD HOUSE

Venus in the 3rd house of the solar return emphasizes all forms of communication, both verbal and written. Your ability to express yourself coherently is important to the activities of the year. Others will listen and respond to what you have to say; therefore your communication patterns need to be clear and concise. For example, one individual organized and wrote a correspondence course during the year Venus was in the 3rd house of the solar return chart. The person's ability to write well was crucial to the success of the task at hand. But one does not have to be involved in a major project to need good communication skills. Effective negotiating techniques are useful in many everyday situations, both business and personal. The ability to make your point quickly, easily, and clearly can give you the edge when mediating a dispute or bargaining for what you want. At other times, being diplomatic may be more advantageous. If you feel the need for improvement, you may want to take a course in communication skills, but for those who are experiencing problems of a personal nature, it is more common to attend counseling sessions, especially when communicating and negotiating needs are the major topics of discussion.

A love of learning is also shown by Venus in the 3rd house. This is a good time to return to school or sign up for a course. If you have been apprehensive in the past about finishing your education or going on for a higher degree, this is a good time to start. It may be easier for you to adjust now since you look forward to learning new things. If you do your best work, academic honors are possible during the year.

Social contacts with neighbors are increasing and you may meet or socialize with a greater number of local people. Superficial familiarity with many neighbors is more common than in-depth relating to a few. The emphasis here is not necessarily on creating new and long-lasting friendships, but acquiring new acquaintances. In keeping with this trend, you may join the neighborhood coffee klatch or community association. My favorite example for a 3rd house solar return Venus is a teenaged girl who took her first job at a local neighborhood library. She not only met many of her neighbors, she also began to date a few. Your urge to meet and communicate with your neighbors can be motivated by a neighborhood problem. Socialization may be secondary to involvement in an organized community effort or public service.

VENUS IN THE 4TH HOUSE

Venus in the 4th house indicates that now more than ever you need a comfortable home to serve as a retreat, a place to regenerate your vitality and nourish your emotional nature. "Home" needs to be a supportive place, a shelter for the wounded, a protection from less hospitable environments. Home may be the place where you are now living, the place you are moving to, or the place where you grew up. It is the sustaining quality that is important, not the location. You need to feel comfortable somewhere on the face of the earth.

If you do have that warm and cozy place to call home, you will enjoy being there this year and may not care to go out much. Rather than nights on the town, you prefer staying at home and entertaining others. If you do not have a sense of home at the present time, it will be your goal to develop a more comfortable living space over the next year. You will want your interests and needs reflected in the place you call home. The changes and decisions you make concerning your external physical environment will reflect your internal emotional changes and decisions.

You will probably redecorate your home. Venus by itself in the 4th house usually indicates simple redecorating. Many planets in the 4th (especially when Pluto, Uranus and/or Saturn are among them) can show major renovations (and a greater tendency toward strong emotional change). If you have outer planets in the 4th with Venus, you may have the urge to redo the whole house, knock down walls or build on additions. Remodeling of this nature is more extensive and tends to take most of the year. During this time you may experience a period of discomfort and physical upheaval as you wait for the renovations to be completed.

There is a reason why the quality of the home life becomes so important. This year can be a time of recovery and healing. Venus in the 4th is often seen in the charts of those who have experienced difficulties in the previous year or years, and who are now focusing their energy on healing old wounds. These individuals need to feel centered and whole, and they focus on their physical home as a source of comfort and rootedness. When Venus is conjunct an outer planet in the 4th house, the healing process is more complicated than simple recovery. Issues need to be resolved and the native may experience continuing or intermittent problems related to the initial wound.

This is a good time to improve your relationship with your parents. If your parents have already passed over, use this year to foster and recall fond memories you may have of them, especially if up until now you have only been able to remember more painful times. If your parents are still living, work to improve your relationship with them. You need that strong sense of rootedness that can come from family ties. If your relationship with your parents has always been difficult and things still don't work out well, this is a good time to form your own roots and fond memories.

An example of the way physical and emotional concepts of home may be intertwined seems appropriate here. A young sculptress found it very difficult to work in her studio. Her father had worked beside her for many years, bonding the wood she carved and doing some of the heavier jobs. He had died the year before and the woman could not bear to enter the studio since it was filled with his presence. She was a sculptress without a home (studio), unable to work. While Venus was in her solar return 4th house, the woman worked with her feelings of grief. She allowed her emotions to surface, remembering her father as he was. She became reconciled to his passing and began to move ahead. Realizing the importance of the external environment, she redecorated her studio to suit her own needs and interests. She added skylights to correct the dim lighting, large windows to enable her to see the trees in the forest, and an oak floor to enhance the natural atmosphere of her surroundings. Incorporated into the studio were things her father had used and she still needed, now placed beside those features that were purely her own. By the end of the year, the sculptress had come to better understand her father and their relationship, the influence of her physical environment and the important role emotions play in her continuing development.

VENUS IN THE 5TH HOUSE

Venus in this house can be indicative of a love affair, especially if the Sun and/or Moon are also in the 5th. Generally, the more planets in the 5th house with Venus, the greater the possibility of romantic involvement. Although any 5th house relationship can lead to marriage eventually, for the present it will probably remain an affair. Strong relationships can be shown by the Sun, Moon and planets in either the 5th or the 7th houses, but marriage is more likely to be considered an option with many 7th house placements. Fifth house placements also imply strong attractions and relationships, but usually the commitment necessary for marriage has not yet been developed. Square aspects between 5th and 7th house planets indicate a disagreement over where the relationship is headed or if it is headed anywhere at all. One partner may be keen on marriage while the other is not. For those who are already married, these aspects and placements suggest that children or outside involvements disrupt your already existing relationship.

Venus in the 5th house does not always mean romance is imminent, because this is also the house of self-expression and greater personality diversification. A positive evolutionary cycle of increasing self-confidence, coupled with increasing self-expression, is associated with this placement. It's a good time to gain confidence in your abilities while trying something totally new. This year can be very creative, productive and encouraging. Self-expression is not limited to artistic endeavors; one individual founded a nursery school with Venus in the 5th house. The medium is not important. What is noteworthy is the growth in self-confidence that accompanies the increased self-expression.

This is a good year to spend extra time with your children and relate to them in a more positive way. If you have had difficulties with your children in the recent past, your relationship may improve during the coming year. Venus taken out of context and by itself can indicate improved conditions according to its house position. Children who have had problems with school, siblings, or adjustments to relocations and social situations, may appear to be making more progress during the year. It is also possible you are better able to understand their problems and take appropriate corrective action. This placement may be mutually beneficial for both parent and child. If you do not have children of your own, you may still be involved with children in one way or another.

VENUS IN THE 6TH HOUSE

You would think that Venus in the 6th house would be a sign of good health, and often this is true. It is possible to experience beneficial changes in wellness, diet and exercise. Healthy individuals can make nutritional changes or establish exercise routines to maximize their energy and physical condition. If you have been ill, Venus may indicate a return to good health, or at least an improvement in your present condition. However, instead of an improvement, Venus can be associated with a tendency towards overindulgence and a susceptibility to its related diseases. The most common problems include, but are not limited to, weight gain, alcoholism, and skin problems. Women can have problems with the female organs. Perhaps the best strategy to use during the year is to practice those health routines which truly make you feel healthy.

The 6th house is also the house of work, and Venus placed here emphasizes the importance of good working relationships and a pleasant work environment. One relationship in particular might be especially important, and you may be romantically involved with a co-worker. All employees will want to have cordial relations with their office mates, while managers will seek to find pleasant ways to motivate their staffs.

Regardless of your position, you can benefit from the use of negotiation techniques, tact and diplomacy in office situations. Your physical work environment is as important as the emotional climate. Your productivity can be increased by beneficial changes in your daily routine. New machinery, such as a computer, may be brought in to facilitate your job; scheduling may be altered to fit your personal needs; nonsmokers may be moved to a smoke-free office; you may be given a raise; or your work could be noticed and appreciated. If Venus is heavily aspected, there will be both positive and negative qualities to your job. Issues will be more complicated, as illustrated by the emphasis in the solar return chart.

VENUS IN THE 7TH HOUSE

A 7th house Venus emphasizes the importance of relating positively to others. This usually applies specifically to your relationship with one particular individual (such as a spouse or business partner) but may apply to others in general. Relationships can improve or deteriorate during the year, depending on your abilities and the situations involved. Effective relating is important to the task at hand, and your ability to foster and sustain good relationships affects your ability to succeed and prosper. For example, if you seek to form a business partnership during the year, your ability to get along with your future partner would have a direct beneficial or detrimental effect on the success of the business.

It is to your advantage to sharpen your negotiating skills and learn to compromise when the situation calls for it, regardless of the type of relationship you are involved in. Realize that you can allow your own personal needs to be superseded by your partner's demands, wishes or expectations. If your spouse is being transferred to another state, he or she may ask you to follow despite your own personal situation or preferences. Your wishes can be superseded by the needs of the relationship, but in the final analysis, the move may prove positive for you both.

This year, you can benefit directly from your associations with others, perhaps more so than if you tried to go it alone. A new writer worked closely with a literary agent to gain the writing contract of her choice while Venus was in the 7th house. She listened and responded to her agent's suggestions about the first draft of the manuscript. Because the two developed a good working relationship based on respect for each other's opinions, the manuscript was improved and the writer ultimately benefited greatly. This is an excellent time to ask for favors and assistance from others. You tend to reap what you have sown.

Negatively, Venus in the 7th house can indicate that you allow a relationship to suppress your individual identity, restrict your personal freedom and tax you unnecessarily. You placate your partner and bend over backwards to please him or her, rather than relate as an equal. You may feel torn between satisfying your own needs and those of the person you are involved with, especially if there are oppositions between Venus and planets in the 1st house.

In an extreme case, the relationship becomes more important than self-preservation and you allow your needs and energy to be drained away. When this occurs, subservience replaces relating, acquiescence replaces compromise, control replaces respect, and you begin to feel that it is your job to keep someone else happy. A young female administrative assistant at a local charity spent more time fulfilling her employer's personal needs than concentrating on her own charity work. She was in a precarious position and since he had the power to fire her, she allowed her own needs to go unmet. Focusing on the cooperative interpretation of this placement will help one avoid the pitfalls.

VENUS IN THE 8TH HOUSE

Relationships will tend to be more intense and complicated during the year. Psychological forces will play a strong role in your feelings of attraction to, and repulsion from, others. You will have a tendency to be unconsciously drawn to certain individuals, perhaps for reasons you find difficult to understand. One relationship in particular may be especially compelling, and you can react impulsively rather than respond rationally to this person. A relationship of this intensity has transforming qualities. It enables you to see your-self, others and relationships in a new way by making everyone's fears, insecurities and complexes more obvious. Psychological com-plexes are impediments to greater intimacy. They must be recognized and dealt with effectively before love can mature. This is an excellent time for joint counseling since the focus for the year will be on understanding how psychological forces play a role in your relation-ships.

Negatively, your relationships, or one in particular, may be a power struggle over money or sex. Regardless of whether this is a love relationship, a business relationship, or a relationship built on animosity, control issues will be important. Either one or both of you may use fear, intimidation, jealousy or manipulation in an attempt to gain money, sex or power over the other. The use of psychological tactics within the relationship complicates and weakens the negotia-tion process, making discussion and compromise difficult, if not impossible. Daily struggles, lacking any foreseeable resolution, can leave you weak, drained, and feeling more controlled than in control.

When Venus is in the 8th house, you may receive money from sources other than your own earning power. If you split funds and expenses with someone else, you can benefit through your partner's pay raise or financial windfall. Your available funds increase through another person's efforts. If you are dependent upon someone for financial support, you may request and/or receive an increased amount of money. Funds can also come to you through an inherit-ance, insurance claims settlement, legal action, or joint financial venture. In all of these cases, money is not work-related unless profit sharing, retirement funds, royalties or disability payments are in-volved.

Negatively, this placement can show a struggle over shared resources. Money intended for your use may not be given freely or may not be given at all. Business partners who cannot settle their monetary disputes may end up in court. Individuals involved in a personal relationship are less likely to go to court over money, but more likely to see money as a tool for control. In this situation, money comes with strings attached. If you are financially dependent on someone else for support or locked into your present standard of living, you are more apt to allow yourself to be controlled, trading your personal freedom for your financial status.

Sex may also be tied to money, power or control. Psychological complexes can color your sexual experience or moral judgments. On a more positive note, sexual fulfillment will be stressed as an important and necessary part of your intimate relationships. Sexual relations themselves can become more pleasurable and better integrated with affection and love during the coming year. This is more likely to be true if sexual problems have occurred in the past and you have actively worked to improve your situation.

VENUS IN THE 9TH HOUSE

With Venus in the 9th house, your beliefs about intimate relationships are being tested. You must support your intimate partnership with a philosophical concept of what that relationship is about and why it exists. Gut reactions and unconscious responses will now seem inappropriate. Relationships are no longer simple boy-meets-girl encounters. They are a crucial part of your present life situation, and issues concerning them need to be examined. One relationship in particular, usually but not necessarily the marriage, must be validated as a worthy and understandable association.

It seems strange that Venus in the 9th house might be associated with relationship difficulties, but this is often the case. Perhaps one tends to dream the impossible dream and yearn for more than is reasonable. If you have this placement, you will naturally tend to review your immediate intimate involvements. You will consider your initial attraction, pains and joys through your years together, and your present feeling of fulfillment, anger or sadness. What you discover may reaffirm your commitment, or make you realize that your relationship is not living up to your expectations. (Expectation, as it is used here, is really a synonym for belief.)

Normally there is a perceived internal or external threat to your main relationship which triggers this review process. Your relationship may not be fulfilling, or you or your partner may be interested in someone else. Sometimes the threat is serious, many times it is not; occasionally it is imagined. It may be only coincidental to the review process. The planets aspecting Venus and the houses they reside in or rule may indicate the issues involved. The main goal for this placement is to establish realistic philosophical and moral guidelines consistent with your intimate experiences and needs. During this period, you might also formulate ethical codes of conduct with regard to professional business partnerships. These guidelines will help you better understand your present involvements while also setting up conduct standards for the future.

As your beliefs evolve, you may need help sorting through the changes. For this reason, many individuals join a religious or philosophical discussion group during the year. The group may not discuss relationships per se, but the discussions will help you focus on the important issues.

You are also more likely to take a course during the year, and the course might be relevant to the issues you are faced with. Educational pursuits may focus on religious or philosophical topics, new-age concepts or relationships themselves. If you attend a formal school, your relationships with professors, students and fellow classmates should be warm and sociable, extending beyond the classroom experience. Benefits derived from your studies may be material, intellectual, emotional or spiritual. Academic awards and recognition are possible for those who do their best work, regardless of whether you are the student or the teacher.

You should travel this year since you will be fascinated by foreign places, cultures and people. Travel overseas may be particularly enjoyable. If you do business with those overseas, foreign partnerships can be pleasant and profitable. If you cannot travel, study foreign customs, try out various ethnic foods and visit ethnic neighborhoods. Romance with a foreigner is possible.

VENUS IN THE 10TH HOUSE

Venus emphasizes the importance of good relationships according to its house placement in the solar return chart. When Venus is in the 10th house, the focus of attention is on relating to authority figures. If you are still young, these authority figures will be your parents and school officials, but if you have reached the age of employment, the interpretation will most likely apply to your employers. This is a good time to foster a personal relationship with your boss; besides making working conditions more pleasant, it can help your career. During the year, who you know will be as important as what you are capable of doing. Your relationship with a boss will probably be strictly business. A few individuals have been able to "sleep" their way to the top, but generally relationships without professional effort or business talent will not get very far.

The greatest success stories come from those who were able to establish a mutually beneficial relationship with their superior whi maintaining high performance. The relationship becomes mutually beneficial when your productivity is enhanced by the personal attention and guidance you receive from your boss, while at the same time your boss's ability to motivate you and other employees is increased by the personal commitment you make to his or her business plan. For some, the mutually beneficial relationship grows to become a mentor-employee involvement which can further your career goals for many years to come, since personal endorsements normally carry a lot of weight. Advancement can come more easily with this Venus placement and this kind of situation. Your employers will be as interested in developing your abilities as you are in getting ahead, so take advantage of this opportunity.

Model yourself after those who are already very successful in your chosen career. Seek career guidance from those you work for. Learn to be a team player in the business world. This is a good time to develop your negotiating skills. Sway others to support your proposals and make compromises work to your advantage.

The 10th house is the house of destiny or long-term influences which become milestones in your personal history. Venus here is associated with making choices that are important to either your career, your destiny, or both. Options are not always career-related, though generally this is the case. You will probably have the opportunity to make an important choice during the coming year. This is a good time to decide to attend college, change professions, become self-employed, relocate to a different state or even a different country. Enterprises you begin now could benefit you greatly in both the near and distant future.

The down side of Venus in the 10th house is the inconsistency of the interpretation. Although Venus can show advancements and success, no planet guarantees a benefic or malefic interpretation. The manifestation is always based on your ability to handle your drives, and the action you ultimately take. This is especially true with Venus in the 10th house; a lot depends on your ability to get along with your superiors. Professional success tends to come more easily with Venus in the 10th, but when no work is involved, it may not come at all. Also, the interpretation for Venus is greatly modified by the aspects made to other planets, more so than is the case with other planets or other houses. Oppositions from the 4th house can indicate that your interests are split between domestic and career needs. Relocation may pull you out of an established, successful position. With square aspects from the 1st house, personality preferences may make you uncomfortable with your present position. You may not like your boss, or you may prefer self-employment to your nine-to-five job. Squares from the 7th house can show conflict between partnership involvements and career demands. Some examples may help you with your own interpretations.

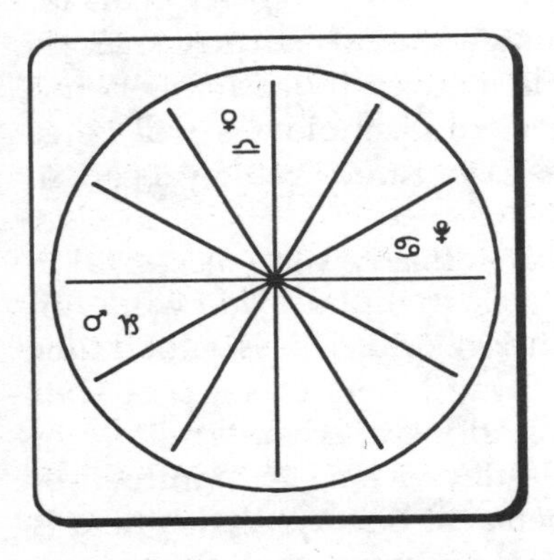

Figure 23

One woman with Venus in the 10th house at the apex of a square with Mars in the 1st and Pluto in the 7th found that her affair with a married man affected her ability to function well at work.

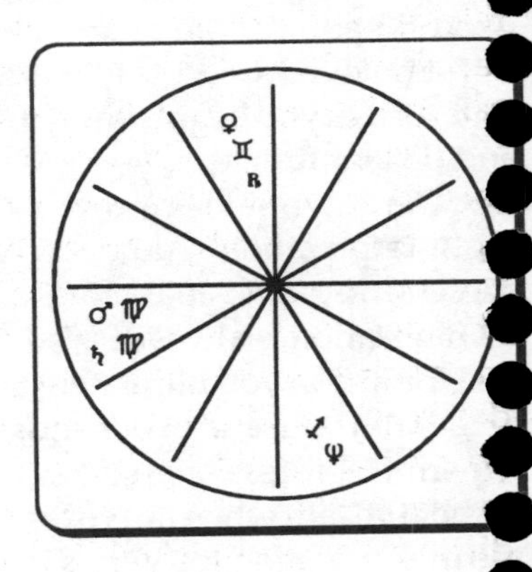

Figure 24

Another woman with retrograde Venus in the 10th opposed Neptune in the 4th and square to Mars and Saturn in the 1st left her job at a design studio and opened her own business. Her ability to be self-disciplined (Mars and Saturn in the 1st house of the solar return chart) enabled her to strike out on her own despite feelings of insecurity (Neptune in the 4th).

VENUS IN THE 11TH HOUSE

An 11th house solar return Venus indicates you are more likely to be involved in friendships and associations with a number of people rather than in-depth relationships with only a few individuals. Your relationship style leans more heavily towards detached, less intimate forms of relating. If you are not presently involved in an intimate relationship, it is unlikely that you will form one during this solar return period. You are more apt to be friendly with a variety of people than fulfilled by any single relationship. This placement can be associated with "playing the field." Even if you do have your eye on one particular individual, he or she may not be ready to make any kind of commitment at this time.

If you are already married or involved in a stable intimate relationship, you may want to make new acquaintances by socializing with groups of people. The focus will be on expanding your circle of friends and creating new interests. Personal freedom may be an issue and you and your partner can make changes which will increase your freedom and flexibility. If your marriage or steady relationship is strong, these changes will not have a disruptive effect. But if your relationship is weak and you question your loyalty to one person or feel very restricted by your present partnership, the push for freedom can lead to strong attractions to other people.

The 11th is the house of "why not?" and issues and conflicts related to monogamy are common. You may be tempted by an opportunity for an affair. How friendly you actually get with others is your decision. Sex is not necessarily a requirement for the variety of experiences and excitement you seek with this placement, but it may suit your needs during this time. Weak marriages can break down, especially if there are oppositions between Venus in the 11th house and planets in the 5th. In this case, friends tend to become lovers while lovers become friends, and your ability to make a distinction becomes clouded. Aspects to the 2nd house show traditional relationships versus the need for freedom, while aspects from the 8th focus on sexuality. Aspects to the other succedent houses can also emphasize monogamy issues, security or financial risks, and stability versus change.

Your friends can be very helpful and great sources of strength. You may depend on your friends to help you achieve your goals during the year. One woman, who started a small business, called on her friends to fill in when she needed extra hands for short periods of time. This is a good year to foster helping friendships and to network within your community. You can do this for either business or personal reasons.

You may join an organized group or attend regularly scheduled social functions. Going to meetings can fulfill two needs, giving you a wide range of social contacts while also expanding your interests. Group involvement can be mutually beneficial, as there is room for give and take with this placement. Interactions can provide you with opportunities for personal growth or professional advancement. You may see group interaction as a way of implementing your goals. This may or may not be the case, but group members are likely to be supportive.

This can be a good time to seek a promotion or raise. The 11th house is the money house for the 10th of career. You may be rewarded for your hard work, but don't expect something for nothing. If Venus is in the 11th without strong Saturnian aspects or placements in the 10th or 11th, the promotion is more likely to involve money only, but with a Saturnian motif, you are more apt to accept extra responsibility along with the salary increase.

VENUS IN THE 12TH HOUSE

When Venus is in the 12th house, it indicates that inner emotional and spiritual qualities are valued more than material success or external achievement. Real success and achievement develop through helping relationships with others and the ability to maintain your own inner peace regardless of your immediate environment or previous involvements. For those who have recently come through a difficult period, this may be a time for inner healing and comfort. Values are centered on nonphysical attributes. Outer beauty and materialism will not be valued as highly as the inner qualities you now seek or possess. Having money is not as important as the quality of life and your ability to be happy with who you are and where you are headed.

The money you already have or presently earn may not concern you; however, a lack of money in the foreseeable future may be important. Venus in the 12th is not especially secure money-wise and it is the future money that is in question. Individuals commonly feel financially uncertain with a 12th house Venus. If you are employed, you may hear rumors of salary cuts, lay-offs or decreasing sales commissions, or you yourself may actually be considering a job change, thereby threatening your financial security. Those who are self-employed may not be able to estimate earnings for one reason or another. Generally a 12th house Venus is more rumor and uncertainty than fact. If you consciously reassess your values and consider your options, you will discover that money is not the focus of your existence; other factors associated with the quality of life are more important.

Because of this internal focus, the spiritual side of your personality can grow and become more intimately involved with your interpretation of daily experiences, thereby making these daily experiences more meaningful and significant within a broader scope of understanding. This is a good time to foster spiritual evolvement through meditation and introspection. "Quiet times" may be essential to your pattern of growth, and consequently you may be socially withdrawn during the year. Being alone gives you time to focus on events, feelings and thoughts without distractions, thus enabling you to complete a process of absorption and integration.

During more sociable periods, your inner focus leads you to be attracted to people, places and things which possess inner strength and significance. Older people, grown wise with their years, and those who seem self-possessed and at peace with themselves are especially attractive to you. "Quality" social contacts are more important than "quantity."

Venus in the 12th house of the solar return also shows possible involvement in a hidden love relationship. New relationships will tend to have a clandestine quality this year. For one reason or another, you find it difficult to be openly attracted to this person. It may be that the two of you are very different and others will disapprove or one of you may have already made a commitment to someone else. Even if the relationship can be more open, the two of you may chose to spend most of your time at home alone with each other.

Involvement in a savior-victim relationship is a real danger with Venus in the 12th house. Since you are more in tune with inner qualities and emotions, you are more likely to be sympathetic and compassionate. You are able to feel the pain and suffering of those you are close to, and consequently you have a strong desire to help.

Your desire to be helpful is especially strong if you are romantically involved. It is to your advantage to help only those who are truly in need, respond to your assistance, and appreciate your help. Unless you are a mental health professional, do not seek to save the alcoholic, the drug addict or the emotionally disturbed. You are not trained to cope with these types of situations and you will be physically drained and emotionally victimized by the resulting frustration. Use your compassion wisely and don't play martyr.

CHAPTER SIX

MARS IN THE SOLAR RETURN CHART

Introduction

The planet Mars represents an outward thrust of energy in a solar return chart. It is an active energy geared towards producing an external manifestation brought about through interchanges with the environment. The arrow on the symbol for Mars illustrates this outward thrust and movement. The interplay with the environment and the area of greatest energy expenditure are shown by the aspects to Mars and its house placement. There can be a number of ways in which the outward thrust of energy and activity manifests in the external environment. Positive manifestations implied by Mars include self-motivation, independent action, initiation of new projects, assertiveness, fulfilling sexual encounters, and original or pioneering creations. But anger, aggression and sexual abuses represent negative and wasteful expenditures. Energy lost in negative exchanges cannot be used productively.

Mars, by its very nature, denotes the spark of initiation and self-motivation. Original pursuits, independent actions and new projects are fostered during the year and are typically associated with those areas of life denoted by the house placement of Mars. Because of the new endeavors and the enthusiasm generated, these areas then become the focus of energy use as the level of activity increases and you begin to feel driven to work, accomplish and complete. Sometimes ambition and competitiveness augment the need to excel, but an energetic attitude can flow without these personality traits since self-motivation tends to spring naturally from an inner source.

During the year, activities may be specifically tied to your ability or inability to be assertive. In its most positive manifestation, the planet Mars denotes an active force. Except for a few of the more subtle interpretations for Mars, the implication is an external thrust which affects other people, places, things or circumstances. For example, if you install a home office (Mars in the 4th), you make a change in your living situation. If you start your own business (Mars in the 10th), you do, in effect, change the environment. If you insist on control over your own finances (Mars in the 2nd), you create a more independent relationship style. In each of these cases, you are the individual initiating the activity and asserting your own needs which are essential to the task at hand. The inability to be assertive can only lead to frustrated desires and anger. Your efforts will be thwarted until you make adjustments in your thinking. When extreme frustration occurs, more energy is expended toward being angry than working towards correcting the problem.

The house position of Mars and its aspects to the other solar return planets provide information about the circumstances which will trigger angry responses in you. Anger may arise from a number of different circumstances, including, but not limited to, frustration at the inability to fulfill your own needs, aggression from others, stressful surroundings or unconscious resentment that has not been resolved. Anger is a signal that your attitude and/or environment need to change. Energy lost to anger cannot be used creatively. Anger that stimulates creative action can be recouped. Your task for the year will be to learn to deal effectively with these anger-producing situations. Do not allow anger to be triggered when insight coupled with assertiveness, tact or corrective action would suit your purpose better. As you focus on negative situations and gain understanding, you can begin to funnel your energies into the most positive manifestations.

Aggression is a distortion of assertiveness. Assertion is the defense and maintenance of your own rights; aggression is the infringement or attack on the rights of others. Anger is most likely the motivation for the attack, although psychological idiosyncrasies may provide other motivations (of which fear is the main culprit). All acts of aggression, by their very nature, should produce anger in the person attacked, though some individuals, because of their own psychological nature, respond in other ways. Where you find Mars by house position indicates where you are most likely to meet the aggressor during the coming year. Either you will have to deal with a hostile person in the environment or you will be confronted with your own acts of aggression. If you are able to meet your own needs through assertion and you know your own rights while conscientiously defending them, you are less likely to be the aggressor or allow aggressive actions to arise in others. You should be able to strike a balance between your needs, and rights, and the needs and rights of others. In this way, you maintain your position relative to the environment. But regardless of how fair you are in your dealings with others, you may still have to defend your own rights.

Mars as an indicator of sexuality is not consistent with the interpretation for every house position. There is the possibility that sexual intrigue can apply to the situations denoted by any of the planet's placements, but the connection is not always obvious. Themes of initiation, high energy output, anger, and aggression appear more consistently. For example, Mars in the 10th may mean a sexual attraction between you and your boss, but it is more likely for you to be extremely active in career endeavors. Your boss can be a prime motivator towards career success or he could actually be very offensive and thwart your efforts. Note the aspects between Mars and your other personal planets for further information. The sexual implications of Mars seem more obvious in the relational houses, but then individual differences apply. Sexual fulfillment is directly related to the appropriateness of sexual activity given the particular situation and any restraints or inhibitions involved.

Planets Aspecting Mars

Aspects to Mars reflect your ability to actively create your environment to suit your own needs and how you would tend to do this. The house placement of Mars shows where the activation is most likely to occur, and the aspects to Mars imply the various internal and external forces created which will either thwart or support your efforts.

MARS-PLUTO

When Mars is in aspect to Pluto, actions are not truly conscious or planned out. There is an acute awareness of the interplay between what appear to be directed activities and unconscious motivations. One does not just set career goals; one is driven to succeed. Compulsions and obsessions, healthy or not, are common since many psychological issues and complexes are indigenous to the scenarios you are involved in during the year. In very negative situations, phobias can develop. The psychological influences affecting you can spring unsolicited from your own unconscious, but are more likely to arise from your encounters with another. Generally, you must deal with this person regularly, and he or she may or may not be totally rational. Reacting from the gut level can become the standard mode of operation for those who do not work toward a greater understanding of these forces.

Control issues are likely during this time, and some individuals get locked into power struggles. In this type of situation, you are both able to manipulate others and subject to manipulation yourself. Surreptitious actions or underhanded maneuvers are also possible. Rather than battling with someone else, you can instead (or also) be locked into a power struggle with yourself. One man with Mars in the 9th house square Pluto conjunct the Ascendent was seriously hurt by a past relationship. He recognized the need to deal with unconscious anger and develop a philosophy for handling future anger-producing situations. During the year, he met and was compulsively drawn to a new relationship. The loss of control over the inhibitions to intimacy frightened him. Consequently, the scene was set and the interplay between the unconscious obsession to resolve the anger issues and fear of being hurt again dominated the involvement for much of the year. Efforts to consciously control yourself will be thwarted until you gain insight into the problem at hand. The man was both irresistibly drawn and frightened by the attraction until he began to resolve issues from the original relationship. This cleared the way for a more meaningful interchange.

Understanding psychological forces and learning to work with them rather than against them can lead to productive encounters. Use insight to break bad habits and negative attitudes. The ability to comprehend new knowledge fostered by the unconscious leads to new power over your own actions and the situations you are involved in. It is at this point that realistic control over behavior begins.

MARS-NEPTUNE

While Mars-Pluto aspects imply actions that are unconsciously motivated, Mars-Neptune aspects indicate actions which have no obvious motivation at all, or which involve a great deal of uncertainty as to direction and goal. Humanitarian pursuits are sometimes associated with this combination since very spiritual endeavors emanate from a source which is not readily apparent. However, the most likely interpretation for this combination is an uncertainty as to future direction. Careful planning is generally not feasible. During the year, you cannot be sure your actions will pan out since there are no guarantees given for your efforts. Job security may be an issue. As an illustration, suppose you wish to run a shelter for injured animals in your backyard and need the approval of the zoning board. You can proceed with your investigations into the matter before the zoning change comes through, but you cannot be sure your efforts will succeed until the final vote. You must trust that situations will work out in your favor.

Mars-Neptune aspects indicate the ability to function despite a degree of uncertainty that colors your actions. If you are correct in your assumptions of eventual success, everything will work out positively; but, if you idealize your actions or situation and miscalculate the results, you will feel disillusioned when you find that your efforts are for naught. Misguided endeavors are possible with Mars-Neptune contacts in the solar return chart, especially with the more difficult or stressful aspects; therefore you must carefully consider your purposes and actions during the year. Be ready with an alternate plan when matters do not progress the way you want them to.

Uncertainty and confusion seem to go hand in hand with this combination and contradictory statements are possible. It is easy for others to misinterpret your actions when your direction is not clear and your plans nonexistent. It is also easy for you to stray from your original purpose when goals are not clearly defined. What you originally set out to do may never materialize. You will be shocked to realize how you have drifted from your own sense of purpose. Use objective feedback from others to keep yourself well-grounded and focused on goals. If you need secrecy for your endeavors, this is a good time to prevent the left hand from knowing what the right hand is doing. But if you need to be consistent, seek constructive criticism from others and don't stray from the path.

MARS-URANUS

Strong changes, usually self-initiated, are associated with Mars-Uranus aspects. Changes range from a constant stream of minor adjustments to dramatic and sweeping transformations. Either form can be beneficial or detrimental depending on individual differences and manifestation. The energy patterns are shifting and the individual no longer wants to be tied down to the same old routine. Something new or exciting is expected and encouraged. During this transitional stage, behavior patterns might be somewhat erratic, and on-again, off-again situations are not uncommon. Changes and the need for greater freedom can lead to sudden separations and broken relationship ties. Freedom of action may be an issue and any restriction will be met with assertiveness if not anger. Speed may be of the essence and therefore time delays are unlikely. Matters tend to move forward quickly.

MARS-SATURN

Mars-Saturn aspects imply a need for well-planned activity which is strongly based on a realistic assessment of the situation. If you are involved in a major project, personality characteristics such as discipline, perseverance, and organizational ability may be needed for the hard work ahead. This is a good time to work towards completing long-term goals, especially if you appreciate the amount of effort that is involved. You are able to accomplish according to your strengths and acquired skills. Future plans are a direct extension of past events, issues and experience. Nothing is given and everything must be earned.

As you strive to succeed in any endeavor, you have to maintain an awareness of societal structures, norms and institutions. Certain requirements will be expected of you as a member of society. For instance, if you wish to borrow money to purchase your first home, you must deal with a bank or mortgage company and meet their requirements for loan customers. This is a good time to push carefully for success, but it is also a good time to cut back on activities if a realistic reassessment of your feelings and life situation tells you to do so. For instance, if you are a young mother wishing to quit work because you have small children, budgeting finances might be a must. Careful planning can make things possible.

Saturn rules reality and practicality. It is not enough to be organized; your plans cannot be implemented if they are unrealistic or impractical. Frustrations are common with this combination and seem to arise from the inability to understand the options and problems. Restrictions and time delays are common in situations where planning is inadequate and expectations unrealistic. Mars by its house placement indicates the action you wish to take (or the anger you feel when thwarted) and Saturn by its house placement indicates where the restrictions are most likely to occur.

Limitations may be built into the situation or personified by a difficult individual. For example, a young man with Saturn in the 4th house bought an older home in need of extensive renovation. Mars was in the 1st house in the solar return and his natural tendency was to go quickly ahead without carefully investigating and planning how to proceed. He soon discovered that problems required more consideration than merely going at it with a saw and hammer. He was able to save time, in the long run, by researching the problem and detailing his plans. Most people would realize that you cannot ad-lib major renovations. When Mars aspects Saturn in your solar return chart, you may have to work with time delays and restraints. In the most negative situations, you could feel like your life is on hold for much of the year. Actions are thwarted by others, if not by your own lack of organization or unrealistic expectations.

Freedom of action can also be limited if liberties have been taken in the past. If you have overspent your budget and have now incurred serious debt, your finances will be restricted during the coming year (Saturn in the 2nd house of money square Mars in the 11th of earning power). You, yourself, may decide to limit your expenses or you may have no other choice but to cut down. If actions are not planned well now or in the past, restrictions, delays and frustrations are a possibility. Those who are able to work within structures will be highly productive.

MARS-JUPITER

Actions which directly support or contradict belief systems are associated with Mars-Jupiter aspects in the solar return chart. Personal ethics and morals may be transgressed or respected, depending on the practicality of the belief, the individual's personality traits and the amount of conflict and stress experienced. This is a time when you are acutely aware of the role beliefs play in controlling or directing behavior. Actions are commonly categorized as right or wrong, while specific relationships are believed to be beneficial and supportive, or detrimental and thwarting.

The danger with this combination is that you can behave in manner which is ultimately not to your benefit, compromising yo belief system through hypocritical actions. Compounding this prob- lem is the tendency to overschedule activities while in the midst o dilemma. You must manage your time in the most effective mann and still make choices which reflect your beliefs and priorities. Long- term goals must be weighed against short-term advantages. T exact interpretation of these issues relates to the positions of Mars a Jupiter in the solar return chart. The action desired or taken is show by the house placement of Mars and the belief system, benefit or ar of heavy activity is implied by the house placement of Jupiter. T aspect between the two signals the relationship existing between th dual messages and whether or not there is a perceived conflict.

For example, one woman realized she could move ahead business by having a personal relationship with her boss. Mars, th action planet, was in the 8th house of the solar return chart sextile Jupiter in the 10th house where she hoped to benefit. She did n perceive the possibility of any long-term problems when initiatin the relationship; she saw an opportunity for advancement. Anoth woman with the same 10th house Jupiter placement had too much a good thing. Her Jupiter was squared by Mars in the 12th house. She was overwhelmed with new business and found it difficult to fi time to work on a long-term project requiring independent resear This woman felt pulled between her immediate success and futu goals. She needed to set priorities. Each of these women fac business issues which might be beneficial or detrimental, and ea made choices accordingly. The task of the Mars-Jupiter aspect is to resolve ethical, moral or philosophical dilemmas, but to do this yo must discriminate between what is an impractical belief, given yo behavior, and what is a hypocritical action, given your beliefs. Fur- thermore, you must prioritize your activities and live within the time restraints you are given.

Mars in the Solar Return Houses

MARS IN THE 1ST HOUSE

This is a year when you seem to have more energy, especially if you do your own thing and enjoy what you are doing. You can get an earlier start, last longer and work harder. For those who are interest in body work, this is a great time for physical exercise. You could ha more energy than you can easily contain and it is to your advantage to work it off in a variety of ways, both physically and intellectually. The energy surge should also be channeled into practical projects. You have the ability to conceive of a plan and independently complete a task. Most likely, the task you choose to work on requires much personal initiative and self-motivation.

This is the year to get in touch with your unique drive for accomplishment. Ego fulfillment and self-interest are important at this time since personal needs are more pressing and demand your attention. You could accomplish a lot if you focus on these needs and allow activities to develop naturally. Much good can eventually arise from your desire to meet and fulfill present and future personal goals. For example, one individual started his own business with Mars in the 1st house conjunct Saturn. He was able to channel his physical energy and organizational ability into a productive accomplishment which required many hours on the job. By the end of the year, he was totally self-sufficient and able to enjoy a very independent lifestyle. Before quitting his former job, he was very angry and frustrated, but the positive use of his feeling of job dissatisfaction led to the new career.

It is true that the seeds sown this year tend to reflect self-interest, but the goal may not be selfish at all. The intent is to apply energy toward self-actualization, and in doing so the ego is able to foster new projects and develop strengths which were previously unrealized.

You should feel good about personal application and hard work as long as you don't overtax yourself and your energy is not drained away by conflict or, by too many diverse interests. Aspects to Mars in the 1st house correlate to interactions and interchanges with various people and situations. Oppositions can indicate power struggles within yourself (yet involving others) as you try to balance your needs with the needs of others. If not understood, these struggles weaken and restrict personal energy. Balance is required. Square aspects can indicate activities which scatter your attention in different directions.

Despite your high level of energy, you can attempt to do too many things at one time. If Mars in the 1st house is squared by Saturn in the 10th house, you might feel pulled between your desire to fulfill personal needs and your need to meet job responsibilities. Time is limited and handling both activities adequately will be difficult. The conflicts you experience with others naturally push you toward functioning on your own (which may be the ultimate goal of this placement), but it is to your advantage to balance all the needs in your life. Failure to recognize personal needs while overemphasizing the needs of others can lead to much anger. Selfish preoccupation can lead to conflict.

Anger in any form is certainly an issue during the year. This is not the time for passivity, and you should be more vocal about what you want. If you feel compelled to defend your rights, you can do this without alienating others. Persistent anger is a negative manifestation and a misuse of energy. You can become your own worst enemy by either abusing others (in the name of anger and self-defense) or allowing yourself to be exploited (self-destruction). In either case, anger drains away your stamina and prevents you from being more productive. It undermines your ability to accomplish. Although your anger is aroused more quickly this year, it is important to handle conflicts effectively. Use assertiveness to take corrective action and then channel your energy in more positive ways.

MARS IN THE 2ND HOUSE

For employed individuals, Mars in the 2nd house usually indicates an income based on commissions or sales. You are able to hustle for a higher income, and your earnings, at least in part, depend on how actively you focus on making money. Even if your income is normally based on a fixed hourly rate, you may be involved in an incentive program or contest which rewards you for the amount of work you are able to generate or handle. If you are considering self-employment on the side, then this might be a good time to piggyback a new budding career on top of a present full/part-time position. Your salary is then augmented by the second job. In some way, the amount of money you earn is at least partially tied to the amount of work you accomplish, and consequently, your income may be affected positively. If you refuse to put in any extra effort or even maintain your present level of involvement, your earnings might be affected negatively.

Mars in the 2nd is also indicative of individuals who either take a pay cut during the year, or do not charge enough for their services. Pay cuts are generally self-initiated, though not always. You can opt for a salary decrease if you really desire fewer working hours and more free time. If your focus is already shifting to other areas, you will not be willing to put extra time and effort into an old routine. Having

More freedom gives you more time to work on future goals you are creating and pursuing. If you are self-employed or directly involved with the future success of your employer's company, you can accept a pay cut to insure the future viability of the business and your job. It is possible that both manifestations of Mars, hustling for money and accepting a pay cut, will operate within the same year. For example, if you start your own business, you may leave a steady job with a good salary. Initially, you earn less and have to hustle to establish your business and draw new customers.

Arguments over finances are possible if you are sharing resources with another person. How you earn money (or what you do for a living), and how you spend money (or what bills you pay, don't pay, and what you choose to purchase), may be points of contention. You feel the need to decide personally how you will earn money and what you will do with it once you get it. For those who are not employed during the year, but are dependent on others for financial support, this placement can be associated with arguments over the distribution of funds. With Mars in the 2nd house, you should normally want to participate actively in the money-making process, but if you do not work, the ambition normally reflected in your own earning power can be projected onto the provider. He or she may or may not have your same desire for success, but generally will not meet your expectations. His or her perceived ineptitude or mismanagement becomes a bone of contention. The underlying issue is not his or her failure as a provider; this person cannot meet your needs since these needs are innately your own. You cannot give away energy and expect others to use it solely for the purpose intended.

Mars in the 2nd indicates your own need to be financially successful, independent and in control of your financial resources. It is difficult to do this when you are not working. The best way to proceed is to take action on your own. Carve out your own little financial niche. Become a fund raiser for a non-profit organization or your child's PTA association. Help your spouse or counterpart by taking the initiative in areas he or she has not considered. Lend support by establishing yourself as an independent and equal financial partner. Don't squander your own abilities and ignore your own needs. Complaints are no substitute for financial potential and ambition. If you feel that you cannot work in any capacity at this time, realize that others cannot and should not be held responsible for your psychological need for success in this area. If you do not understand this principle, your behavior can become self-defeating. You can spend money impulsively, faster than your provider can make it, which is a negative form of financial control (or lack thereof).

The 2nd house is also the house of self-worth, material values and traditional morals. During the year, you will be very aware of how others treat you and whether or not you receive the respect you deserve. If you are not fully appreciated, you can assertively or even aggressively stand up for yourself, your decisions and your lifestyle. Your material value system is changing and you set new priorities which are more consistent with your personal perspective. Moral decisions are made independently and you are willing to support your stance vocally. You are not afraid to go against the grain and will actively pursue what you deem worthy or essential even if you must defend or protect yourself from the remarks of others.

MARS IN THE 3RD HOUSE

During the coming year, you have the ability to speak with great urgency and can motivate others to respond to problems and solutions as you see them. Your speech may not be elegant and memorable quotes are unlikely, but your main goal is to incite others to action. You are assertive, direct and to the point. In the assessment of any situation or dilemma, you are able to cut to the heart of the matter and concisely state the issues. If you must mediate a dispute, you will not pussyfoot around sensitive topics. Your mind is sharp and you will call a spade a spade. Your insights lead to action and your negotiation tactics will not include placation or appeasement.

Ideas move quickly through your head and therefore this is an excellent time for learning. A high level of energy and enthusiasm can permeate your studies and you can grasp central concepts with ease. But, if you allow your mind to wander aimlessly and scatter your thoughts superficially through many topics, knowledge will be based only on half-truths. There is a tendency to scratch the surface only and not follow through. Impulsive decisions based on incomplete information are possible if you do not wait patiently for all the facts to be made available to you. You can jump to conclusions and respond without thinking, especially if you are angry or unconsciously motivated. Mars in the 3rd is not generally associated with accurate thought processes, but more likely rapid thoughts, and unfortunately accuracy can be sacrificed for speed.

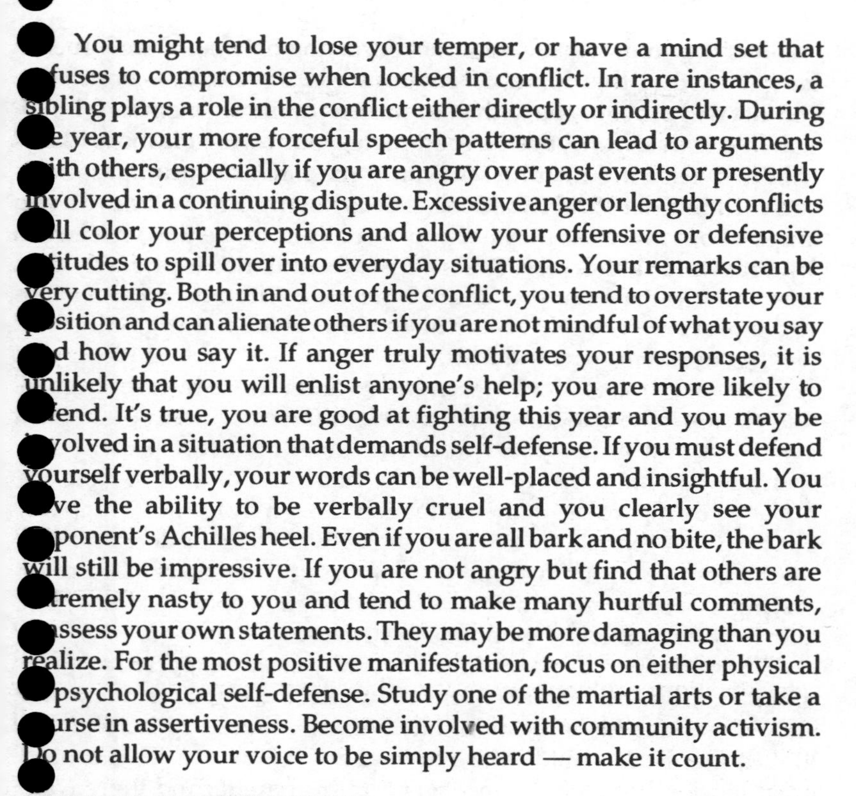

You might tend to lose your temper, or have a mind set that refuses to compromise when locked in conflict. In rare instances, a sibling plays a role in the conflict either directly or indirectly. During the year, your more forceful speech patterns can lead to arguments with others, especially if you are angry over past events or presently involved in a continuing dispute. Excessive anger or lengthy conflicts will color your perceptions and allow your offensive or defensive attitudes to spill over into everyday situations. Your remarks can be very cutting. Both in and out of the conflict, you tend to overstate your position and can alienate others if you are not mindful of what you say and how you say it. If anger truly motivates your responses, it is unlikely that you will enlist anyone's help; you are more likely to offend. It's true, you are good at fighting this year and you may be involved in a situation that demands self-defense. If you must defend yourself verbally, your words can be well-placed and insightful. You have the ability to be verbally cruel and you clearly see your opponent's Achilles heel. Even if you are all bark and no bite, the bark will still be impressive. If you are not angry but find that others are extremely nasty to you and tend to make many hurtful comments, reassess your own statements. They may be more damaging than you realize. For the most positive manifestation, focus on either physical or psychological self-defense. Study one of the martial arts or take a course in assertiveness. Become involved with community activism. Do not allow your voice to be simply heard — make it count.

MARS IN THE 4TH HOUSE

You are more likely to expend energy working around the house. You can redecorate or renovate your living quarters, but projects you decide to tackle will probably involve lots of do-it-yourself elbow grease. Renovations are not necessarily beautiful, but do tend to involve a lot of hard work or at least more work than would normally be required. Repairs and renovations tend to be "first time" projects. For example, a newly married couple may inherit a lot of used furniture from relatives which they decide to refinish before placing in their apartment or home. Consequently, the decorating project becomes more difficult than simply purchasing furniture. Minor repairs tend to grow into major hassles that require a lot of work once you get into them. Simply retiling a bathroom is hard enough, but when you discover you must replace the rotten walls underneath the tile, the job becomes much more demanding.

It is common to have a conflict occurring between family members or within the domestic environment. If you live alone, the conflict will probably be with your parents or family members, but if you live with anyone, even a roommate, your conflict is more likely to be in the home. Sometimes the confrontations can encompass both sets of individuals and situations. It does not matter whether you are angry about events as they occur or situations in the past; the anger is just as real in either case. The reality of the situation is probably that past and present perspectives are connected. Circumstances similar to the past are probably occurring in the present and triggering old feelings of resentment and hurt.

This is a time to air old resentments and release hurt feelings so that circumstances in the future can be different. Unless you understand the past hurts, you will not be able to correct present-day detrimental situations. It is possible that those closest to you are hurting you regardless of their intention. They may not see your pain, but you can explain it to them logically if you understand the mechanism yourself. Perhaps you are being taken advantage of or staying in a situation which is not emotionally healthy for you, but allows your conflicting emotions to surface.

For example, a teenager from a wealthy family expected to go on to college, but his parents were divorced and fighting over who should pay for his education. Neither one offered to help and it looked like both would refuse to lend any assistance. This caused great tension between the teenager and his parents and there were many fights in the home. The teenager honestly felt that he had raised himself to begin with, and had allowed his parents to take advantage of his willingness to accept responsibility for himself even at a very young age. It was during this Mars placement that he realized he had allowed an emotionally unhealthy situation to persist. All during his childhood he had managed alone and never had the luxury of being able to depend on his parents. He also never confronted his parents or his anger with them. He was now very upset that his emotional needs were never met either in the past nor in the present. He became very vocal about this specific pattern of relating, but less argumentative over little things. He zeroed in on the root of the problem and subsequently sought out a family counselor.

This is a time when unconscious anger surfaces and individuals begin to see behavioral patterns reminiscent of the past and most particularly childhood. Seeing these patterns creates great stress, but the goal is not to live with the conflict, but to break the pattern. The need for emotional self-protection is implied by this placement, and you should want to protect yourself from future uncomfortable situations. It is up to every individual to seek new ways of interacting which fulfill rather than compromise emotional needs. Try to understand the "why" of anger as well as the "when" and "how." Repetitive conflicts which lack any insight allow you to spin your wheels without growth.

It is particularly important to understand family interactions since they are the key to all relationship dynamics. If Mars is in aspect to Neptune, conflicts may be built on misunderstandings and you should investigate. Once a divorcee with small children chose to live far away from her parents. She honestly thought that they would not want her around since they never offered to help her during the divorce proceedings. But her parents were very upset that she moved away and never sought their help. They assumed she knew she was always welcome in their home and they were ready to lend a hand. They held back, not wanting to interfere, while she left, not wanting to ask for help. Neptune was in the 10th house opposed to Mars in the solar return 4th house. This is a good time to take the initiative in handling emotional conflicts. If misunderstandings are occurring, open up the lines of communication. Seek counseling for very difficult and detrimental situations, and protect yourself from further stress. Use assertiveness rather than anger to dismantle unconscious complexes which lead to painful interactions.

A few may decide to set up a home office and work totally or partially out of their residence during the year. Mars can be indicative of a side-line business you start or enter into with other family members. If you retain your regular job, it may ease your workload if you can bring more work home with you in the evening. Adjustments or conflicts in the balance of work and homelife can be an issue. This is true both for those who work out of the home and those who bring work home. It is also true for individuals or couples sharing household chores and repairs. The division of labor within the home can be a topic of discussion or conflict. Renegotiating responsibilities for family chores commonly occurs when life situations are changing and one individual is now home more or less than previously.

MARS IN THE 5TH HOUSE

Mars in the 5th house indicates a more assertive style of self-expression. This may be most evident in creative projects, but it can also manifest in the personality. You are less likely to practice self-restraint and more likely to take risks. You are willing to go out on a limb with a new train of thought. You gain confidence by trying new things, although in the beginning your confidence tends to be fragile. This is a time when you can state what you want and go after what you need. You are direct and concise with a no-nonsense style that lets you be who you want to be. You refuse to surrender your right to freedom of self-expression, and if someone is attempting to suppress your ideas or restrict your movements (regardless of their motives), you will fight back. For example, an astrology student with this placement consistently refused to give up her studies when her fundamentalist parents objected. Despite their feelings, she could not be swayed. You are very aware of the personality's need for self-preservation and integrity. If someone is a threat, malicious or otherwise, you are ready to defend yourself. If you are involved in a continuing conflict, you will be willing to take the offensive if the situation warrants it. Some may think you are too pushy, and perhaps you are if you need to be reminded of the rights of others. Use this energy for creative endeavors and free self-expression. Do not focus on dominating or controlling others (especially children and lovers).

It is not especially common to start a love relationship during the year, but if you do, the relationship can be very exciting and based on a strong sexual attraction. Desire and passion are most likely evoked by the person you are seeing, and these feelings play an important role in the relationship choices you make. Romance may not be as important as the electrifying energy that passes between the two of you when you are together. But there is no guarantee that a relationship will be all passion and no conflict; in fact, the fire that heats up your passion can also heat up your temper. In addition to the strong attraction, there is also a need for self-defense and unrestricted self-expression as discussed above. How do you defend yourself against a strong passion which is essentially an invisible power difficult to control? How do you maintain unrestricted self-expression when you are strongly involved with another person? Conflicts in relationships are common since couples will feel both the pull of attraction and the push for expression, or the desire for merger balanced by the need for freedom. Learning to maintain some measure of self-control and self-expression while in a relationship is the key to this placement.

If Mars is strongly aspected in the solar return chart, the relationship will be further complicated by other needs and issues in your life. By example, if Mars is squared by a planet in the 2nd house and you are presently married with traditional monogamous morals, guilt can conflict with your passionate involvement. Mars opposed from the 11th can indicate a relationship which vacillates in its intensity. You and your partner may wonder whether you two are really lovers or merely friends. If a planet in the 8th house is square to Mars in the 5th, you and your lover may have different sexual rhythms, preferences or ethics which need to be discussed and balanced.

If you are a parent, your children may be pushing for new levels of independence, regardless of their age. One-year-olds go from crawling to running, toddlers enter the terrible twos, older children start school, teenagers learn to drive, get that first job, leave home to attend college, or start living on their own. The task they are working on is not as important as the reaction it causes in you. Basically, you are not ready for the changes and may disapprove of the choices they are making or the behavior they are exhibiting while out of your sight. Children appear more disruptive during this period of transition, but you are also less patient. They are probably coping with new situations, abilities and problems. The newness builds stress and they are easily frustrated. You, in turn, wish they would return to their old pattern of behavior which seems more settled and less taxing.

If your children are younger, they may disobey you, but if they are older, they will not only disagree with you, they will argue with you. They are probably intelligent enough to point out all the flaws and discrepancies in your restrictions and disciplinary tactics. Don't be surprised when they confront you with a logical attack on these inconsistencies. This exchange forces you to further define your position on any one issue and enables the child to push for change and freedom consistent with his or her real or imagined level of maturity. You, however, may not see it that way and attribute most of the tension to simple parent-child confrontations. These minor conflicts tend to be aggravating, but easier to handle than the well-planned, persistently orchestrated debate put on by some teenagers.

For those more creative individuals, this is a time for original work in new areas of expertise. You should be willing to take artistic risks with your creations. If you are an artisan, experiment with new products or designs. If you are a writer, try a new style or genre. Innovations are important, and you can receive constructive criticism over your new approach. You should be ready to defend or explain your choices, but also use the information to refine your technique.

MARS IN THE 6TH HOUSE

The ability to work independently or with a limited amount of supervision is shown by Mars in the 6th house. You can work entirely on your own or set up your own business, but independence can be relative. A medical laboratory student completed a year-long internship with intense supervision and graduated to her first full-time job when Mars entered this house. During the coming year, her work no longer required constant verification; however, she continued to work in a large laboratory setting with others. If possible, it is better for you to work alone and/or be your own boss. You have the ability to be self-motivated and can initiate and complete projects without prompting.

Workaholic tendencies are likely, especially if you have deadlines to meet or if crises periodically occur. Compulsive work habits that are carried too far eventually begin to affect one's health negatively. On-the-job tension is possible for those who work too hard or find the working environment unpleasant. You could be easily angered or frustrated by your working conditions, co-workers or employer. Conflicts can ensue. A fellow who was used to a very professional environment found it difficult to adjust to the lackadaisical attitude of co-workers and managers when he changed jobs and began working at a new office. He was aghast to discover the poor quality of service being rendered to clients; consequently, the situation caused him great stress. If you are easily angered and frustrated by your work, make suggestions which will help to improve conditions for everyone. Complaining is associated with this placement, but it would be more advantageous to take the initiative for problem solving. If presented tactfully, your ideas can be accepted and may help to straighten out office difficulties or conflicts.

Caution is warranted when Mars is in the 6th house of health. If your health is slipping, start to reassess your habits, specifically looking for health practices or routines which are not good for you. It is not uncommon to find you are working against your own well-being. If you are presently healthy and follow a beneficial program of exercise and diet, you need not have any difficulty with this placement. Problems are more apparent when you are doing something you should not be doing. Generally, negative patterns do not develop during the year, but only become more obvious. This is the time you begin to see the effect of these detrimental habits and possibly feel pressured to control your behavior. Basically healthy individuals are more likely to discover minor impediments which can easily be corrected. For example, one woman noticed that her cholesterol count was rising and realized she consistently ate foods high in fat content. Another realized she was overdosing her body with megavitamins.

Common minor problems include overeating, harmful dieting, consuming foods which upset the system (such as caffeine products, hot or spicy dishes, salty snacks, allergy-producing substances), failure to treat minor or chronic health problems, over-scheduling activities, and aggravating skeletal problems with strenuous activity or lifting.

Serious health problems are related to self-destructive habits which tend to be especially noticeable and threatening this year. The more serious threats to health correlate with the more excessive habits practiced over a long period of time (obesity; drug, alcohol or cigarette addictions; self-neglect or abuse; compulsive or stressful work habits; anorexia, bulimia, malnutrition, yo-yo dieting or other indiscretions). Surgery is possible during the year, especially those surgeries associated with infections (appendectomy and tonsillectomy), and those surgeries considered to be elective (sterilization, cosmetic, or corrective).

The purpose of this placement is to make you more aware of any detrimental habits you might have. Correcting the habit can alleviate the health problem. Learn to take better care of yourself. This is an excellent time to exercise your way to better health.

MARS IN THE 7TH HOUSE

This is a time to be energized by another person who can get you to do something you would not or could not do alone. In its most positive manifestation, collaboration with another carries you towards the initiation and completion of a project or joint endeavor. If the two of you can agree on mutual goals, you both can make great strides together. Your partner's enthusiasm and drive become the primary motivating force. Most likely, it is the other person's character which provides the spark of inspiration and the encouragement necessary to begin the task, but it is the interactive qualities of the relationship which keep each of you at a high energy level, working towards completion.

Hopefully, you will be dealing with someone who has your best interests at heart. The problem associated with this placement is that you can be led down the wrong path if you do not choose your friends and partners carefully. If you are connected to individuals who are bad for you, their effect will be negative. For example, if you wish to start a food cooperative and your spouse feels you are too inept to succeed, his or her negative influence will hurt your chances for success. But if he or she is totally supportive, and in fact, leads the way, your chances are greatly improved. A lot depends on the partner's attitude. Since you are so susceptible to the influences of others, it is important to assess who the others are. Contacts can range from the most positive to the most detrimental, from successful achievement to damaging encounters. In the worst manifestation, involvement in a crime is possible. It is important to identify those relationships which are wrong for you and counteract their influence. Cultivate those more positive involvements which enhance your psychological health and soul qualities.

There is the possibility that you will be in a relationship which involves conflict. You may be angry with someone you see daily or this other person may be angry with you; anger in both directions is possible. It seems that every time you try to express your feelings to this other person, the two of you become involved in an argument. If this is your present situation, you could be involved in an ongoing conflict that continues throughout most of the year and colors your perceptions and attitudes.

Eventually, your frustration can spill over into other areas of your life and affect other relationships as well if you do not understand and contain the negative impact the conflict is having on your life. Disagreements may be poorly defined or specific, but generally involve a daily ongoing process. For example, a young man was to have inherited his grandfather's watch, but unfortunately, just before the grandfather died, another relative had him change his will and the watch went to an uncle instead. It just so happened that the young man worked with the uncle daily in a family-run business. The uncle spent much time talking about his good fortune. The young man constantly felt angry and tense at work because of the uncle's presence and actions.

The conflict may only involve a general dislike for a person you see on a regular basis, or it can be more. Interactions range from mild irritation to legal battles, personal vendettas, both sexual and nonsexual harassment, and aggression of any kind. More than likely, another is an aggressor if you allow yourself to be the victim. Don't lose sight of your self-esteem and need for self-expression. Be aware of your rights in any given situation. Someone may be telling you what to do, what to wear and who to be. You need to learn to practice assertiveness this year and it could be necessary to learn to protect or defend yourself. In very difficult situations, you need to decide whether you will fight or retreat. Some situations can be ignored, or controlled; others require a more definitive response.

Learning to cope with angry people and angry situations may be part of the lesson associated with Mars in the 7th house of the solar return chart. Now is the time to deal with unresolved issues even if they have lain dormant for many years. Time is running out and you must act now. Learn to handle the curt responses and hurtful remarks and go for the heart of the conflict. Do not get sidetracked. Force the issues into the open so they can be discussed and understood from all sides. As long as anger remains denied or insidious, problems cannot be resolved. If resolutions are totally unattainable, learn to shield yourself from the barbs or retreat to more beneficial situations. Do not stay in situations which are detrimental or counteract positive growth and insight.

Mars is also the planet of competitiveness. Here in the 7th house, it shows that the spirit of competition spurs you to action. You will be motivated to work harder if you are competing with peers or even enemies. The "I'll show you" attitude can work in your favor. You are very sensitive to criticism from others this year, and anticipation of future criticism, either real or imagined, can motivate you to work harder now and get more done.

MARS IN THE 8TH HOUSE

There is a transforming quality associated with Mars in the 8th and much energy can be applied towards gaining insight into sexual, financial and psychological matters. Resulting changes may come through enlightenment or through conflict. Either way, changes are likely to be very strong and powerful.

Sexual issues revolve around changing sexual needs, attitudes and ethics which are in a state of transition, possibly affecting others.

During this time, physical needs may increase or decrease and sexual practices can change to meet those needs accordingly. Owning your own body, recognizing your sexual desires and taking responsibility for your own actions can be an important process leading to mature sexual ethics. The quality of any sexual relationship is particularly important and you will become aware of the full range of expression either in your own life or through contacts with friends, neighbors, family or casual acquaintances. Exchanging sexual information will help you weigh your experiences and can be crucial to further enlightenment and understanding. Should you be involved in a fulfilling sexual relationship, the transforming quality of your involvement will augment your growth and positively influence other areas of your psyche. Should you feel sexually unrewarded, used or abused, the physical demands on the body and the psychological damage to the psyche will impede growth.

Very negative situations exist when partners struggle over sexuality and ethics, or end up exchanging sex for money, position or control. Purely sexual conflicts focus on the differences in moral standards. You, yourself, can be of two minds concerning your own sexual behavior, or you may conflict with those you are emotionally involved with. Sexual difficulties between couples (including conflicting needs, schedules or idiosyncrasies) are possible when Mars is very stressed. Aspects from the 8th house Mars to any other succedent house planet can illuminate the characteristics of these dilemmas. Sometimes the conflict involves both sex and money. For example, married couples who have built their relations around conflict and a persistent lack of cooperation may exchange money for sex. If the wife has sex with her husband tonight, he gives her money tomorrow. Other, more subtle variations of this scenario can occur, such as a business associate who uses sexuality to sleep his or her way to the top.

Financial issues related to Mars in the 8th house usually reflect the partner's need to control his or her own resources and the possibility of cost-cutting measures. Concern over debts is likely. If the debt situation is severe, this is a good time to see a financial advisor. Disagreements or even fights over spending practices and joint resources are possible. The tendency will be to haggle over "who pays for what," but major conflicts need not ensue. Separated or divorced partners might fight over alimony or child support, and the conflict can end up in court, especially if the litigants are stubborn.

and oppositions run between 2nd and 8th house planets in the solar return chart. Contested wills and inheritance conflicts are also possible.

Psychologically, Mars in the 8th house can be indicative of deep-seated anger and long-lasting resentment. You might be upset with someone for transgressions that have occurred in the past or conflicts that are occurring in the present. If your disagreement was in the past, Mars here shows the resentment you still hold and need to resolve. Until you do release this negative energy, you will feel as if you are at war with yourself. Pervasive anger blocks you from achieving your fullest potential. This is a time meant for reviewing old resentments and defining exactly what you were angry about. Understanding past situations in the light of new information and more mature attitudes can help you release psychological blockages. Confronting anger intelligently can lead to greater freedom.

If you are involved in a present-day conflict, you can be locked into a never-ending battle with another person if neither of you will agree to a compromise. This conflict can be detrimental to your mental health if it involves psychological manipulation. Guilt and fear are the main manipulative tools used to control the psyche and you may be the victim or the perpetrator. Do not allow manipulative people to control your behavior by pushing your buttons. Dismantle those responses that are not in your best interests, and deal intelligently with manipulations, controlling messages and power issues.

Regardless of whether the conflict is a past or a present one, your anger will not fade easily until it is examined and understood. Learning to gain control will be an important task for the coming year and the more you understand about your own anger response mechanism, the greater the power you will maintain over your destiny.

MARS IN THE 9TH HOUSE

This year you will identify strongly with your religious, philosophical and ethical beliefs, and this is a good time to actively pursue further investigations in these areas. Round-table discussions of these topics lead to a clearer understanding of issues and possible implications. Spiritual concerns and dilemmas may be especially important during the coming year. Learn to share your thoughts rather than force issues. Support your words with actions and work for those causes you believe in. Motivate others with the strength of your conviction. Your commitment can be inspirational. Positively, this is a time when you can become a strong proponent of truth and justice; but, negatively, you may be prejudiced or intolerant of the beliefs of others if you use your convictions to assume a holier-than-thou stance. There is no guarantee your beliefs will be beneficial to your spiritual growth or to those you come in contact with.

The 9th house is the house of all beliefs, not just those that are philosophical in nature. With Mars in this house, if you believe you have been wronged in the past, you will tend to take an offensive rather than neutral position regarding similar issues, events or relationships. Anger could be inherent to your belief system, making you feel duty-bound to correct past situations or eliminate the possibility of recurrence. If you work in the humanitarian field, you can fight for the rights of underprivileged people rather than brood over a personal problem. Mars in the 9th is noted for soapbox warfare (standing on a soap box and pleading with the masses to end this deplorable situation). But the difficulty with this placement is the very negative manifestation which allows the individual to justify offensiveness in personal matters. Inequities in the past should not become an excuse for inequities in the present. A strong defense is more appropriate to this placement than a strong offense (as long as the defense is not an impenetrable fortress).

During the year, review your responses to anger-producing situations and consider ways to express anger in a more constructive manner. You need to learn to cope effectively with certain common or recurring situations in your life. You may not be able to control these difficult situations completely, but you are able to control your reactions. Choose a course of action which provides you with a good response while not escalating the problem. Now is the time to develop a game plan for handling anger-producing events, especially if you presently find it difficult to cope. Your anger might have been ineffectual in the past. Jumping up and down and screaming is generally ineffective. So is total retreat. Learning to express anger intelligently is an important task to be mastered during the year.

The 9th house is also the house of travel. With Mars in this house you tend to take quick and perhaps nerve-wracking trips. Most likely, you may not be able to relax while traveling, and problems with weather, travel connections, and accommodations are possible. This does not mean that you can't enjoy traveling, but you should be careful not to pack too much into each trip. If you are very careless or angry, it is possible to have a minor accident while traveling, but this will probably occur only in very difficult, careless or angry situations.

The mind can be very active, and consequently this is a good time to be self-taught. You can motivate yourself to gain knowledge while studying on your own. You should be involved in some intellectual pursuit, either alone or with others, be it school, teaching, lecturing, learning, etc. This is not the year to let your mind lie idle. Do research and find your own truth. Actively acquire information and organize it according to your beliefs.

MARS IN THE 10TH HOUSE

Mars in the 10th house suggests an aggressive business style or energetic attitude towards career tasks. Many times the individual with this placement realizes that he or she must work harder during the coming year either to get ahead or stay ahead. Competition from other businesses or co-workers can provide the impetus towards greater productivity, but many times the individual is only in competition with himself or herself. There may or may not be a specific change in the work environment that indicates this need to switch into high gear. Internal signals are as likely as external motivations.

This is a good time to focus on career ambitions and use the energy surge for both immediate career goals and long-term developments. During the year, you should take the initiative and capitalize on your ability to be self-motivating. Projects that require originality, independent work and/or aggressive action are suited to your style. Try new techniques, tools and directions. Use this time to initiate a project or start your own business. Professional success can result from independent efforts, and as a rule, you will prefer working alone. Specific situations which are consistent with this interpretation include: working on a commission basis and trying to land the "big deal"; attracting new clients, particularly those requiring active pursuit; and starting a new business or relocating an old one to a new area, since you must hustle to make things work, break even or move ahead.

Those who cannot find a positive outlet for this energy surge will feel frustrated with their present employer or career responsibilities. If you are not self-employed, or do not have the freedom to work independently, clashes with authority figures are likely. In some instances you can be motivated by your boss's demands, but generally, tension between you and upper management can build into a stressful situation for one reason or another. High pressure situations and workaholic tendencies are associated with this placement and if your boss is too demanding, your workload can be overwhelming or even impossible. You can accomplish a lot this year, but do not do so at the expense of your mental or physical health. (Check aspects going to the 6th house.) Do not allow workaholic tendencies to prevent you from going on vacation. It is important that you practice relaxation techniques and take time off. It is also important to learn to control, deflect or address tension-producing issues.

In very negative situations, the stress at work is exacerbated by unreasonable or difficult authority figures. It would be best if you could work alone, since you may not like working for, with, or under anyone. The natural creative energies will be more evident if the motivation is internalized and allowed to flourish, but this is not always the case. If you must deal daily with an argumentative boss, you need to look at your professional priorities and reassess your options.

In the most positive manifestation, energy is expended towards professional achievement, but if the road to success is blocked, extreme frustration will follow. Making job or attitude changes can be the only way to open doors to more positive opportunities.

Although the 10th house is associated with career choices, this is also the house of the destiny path. New personal directions are possible while Mars is placed here, regardless of whether or not these new directions are directly tied into career choices. Occasionally this placement shows that you are able to go on in life alone. This is particular true for those individuals in the midst of a divorce. The ability to function independently while pursuing a course of action which can have a major effect on the life path is an interpretation consistent with Mars in the 10th house of the solar return chart.

MARS IN THE 11TH HOUSE

Mars in the 11th house shows an intense desire for freedom of action and the aggressive pursuit of goals. You will not allow or accept limitations on options. You do not want your present short-term goals or your future long-term goals to be dominated, controlled or restricted by others. You will protect your right to proceed in any area of personal or professional endeavor. The goals you set during this year can generally be accomplished without the support or assistance of others and despite any objections. To accomplish your goals, you may have to bend a few rules or break a few family traditions. You are not afraid of independent action and it may be necessary for the task at hand. For some individuals, goals will be very self-oriented and self-serving, perhaps even selfish.

For example, a woman about forty years old had worked at her husband's side in a family-run business for twenty years. When Mars was in the 11th house, she discovered a great love for another professional field while pursuing a hobby. She decided to quit her current job, but her husband objected, saying he needed her to manage the office. Eventually she did quit despite his objections and began working for herself. She initiated the business on her own despite the family objections and continued to pursue a goal only she was interested in.

The goals you seek during the coming year will normally be in keeping with your personal needs and abilities and are the product of your own mind and not an offshoot of someone else's ambitions. This is a time to feel both self-sufficient and self-possessed. External manifestations are meant to reflect inner qualities. The active creation of dreams will help you to discover your own ability to function independently. Your range of future activities will no longer be restricted by a need for help or agreement from others. You are capable of initiating things on your own and following through. Focus on your own unique individuality and value your abilities, whatever they may be. Develop and pursue talents and ideas that may have been lying suppressed for many years. This is a time meant for reviving those dreams lost when they seemed, at the time, too selfish to pursue. You can now accomplish projects abandoned in deference to others, but you must be assertive and ambitious in your desire for fulfillment.

From the perspective of others, family and friends, you may be somewhat unpredictable and undependable. Others may be concerned about your newfound independence, but assurances should be enough to calm their fears if you maintain a measure of stability. More difficult is the total disregard you might have for the needs of others and your responsibilities to them. You are capable of making decisions impulsively and without consultation. You are also capable of forming goals without adequate research or reflection. Some individuals will truly not handle their new-found freedom wisely, and consequently, they become the cause of great concern to others

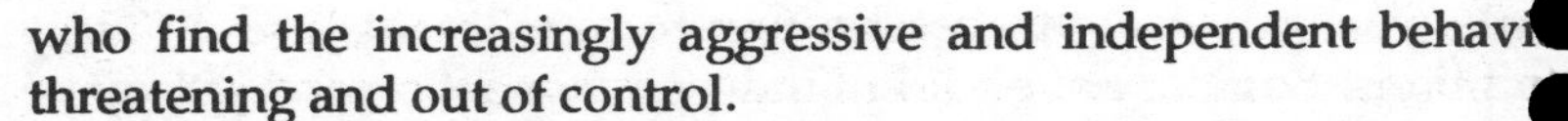

who find the increasingly aggressive and independent behavi threatening and out of control.

Rather than choosing to work alone, you may become involv in a group effort or work on a project with a number of your frien But even in this situation, you will most likely gravitate towar leadership positions. You can be a great motivator for others and yo yourself, will also be energized in return. If you act as an independe within the group setting, others will resent your preoccupation wi personal needs or fulfillment. Conflicts within the group are possibl and you can be drawn into any power struggles.

Conflicts with one or more friends might also occur. For examp a young college woman was involved in a very difficult friendsh with two close friends who were using drugs. Eventually, the wom withdrew from her friends because of the drug use. She returned her studies and went on alone. During the year, you and a friend ma be at cross-purposes and this will cause contention, but you c instead be motivated by common interests and shared goals.

MARS IN THE 12TH HOUSE

While Mars is in the 12th house you have the ability to wo independently or behind the scenes. This is a great time to focus on projects which require some degree of "aloneness." You must p vide your own sense of direction and motivation while worki towards achievement, but what you ultimately accomplish can be very original and unique. Year-long projects are associated with t placement since many people will not know what you are worki on or have accomplished until late in the solar return year. The tendency with this placement is to prevent the left hand from kno ing what the right hand is doing. It is even less likely that others w know what you are doing. This is your time to work in secret or "go undercover." Suppose you wish to surprise your spouse parents with some secret home-made project you assembled in t basement; now is the time.

This is not your year to be openly aggressive or angry. You te to check your temper and bite your tongue, ignoring even offensi remarks from others. Your reflexes are slow and therefore you miss your chance to respond. More than likely, you think of a go comeback hours later. Occasionally, you will realize the next day th you should have felt insulted the day before. You do not have adequate or timely defense tactics, and even anger itself can delayed or vague. For these reasons, you are more likely to withdraw from confrontations, especially those with family members. You may still feel angry, but you are less likely to express your anger openly and will not be as argumentative as you have been in the past. If you do not correctly understand your responses and your situation, pent-up emotions could cause you to feel very stressed and irritable. As the tension builds, gut reactions will tend to take the place of informed responses. Distortions in reality perception can occur. Actions and words will be inappropriate for the immediate situation if you are responding now to something that happened yesterday. If you allow tension to build to this extent, suppressed anger can cause you to act in a way that is counterproductive to what you hope to accomplish. In very negative situations, anger is displaced, moving from the truly annoying person to a less threatening adversary or innocent victim.

Extreme anxiety during this year is ordinarily associated with the inability to express anger or defend oneself in difficult situations. The inability to understand what is happening psychologically contributes to the nervousness. Specific situations and people are most likely to trigger the undue stress. Normally, anger and a feeling of defenselessness underlie all anxiety.

In many cases, there will be logical reasons for your lack of aggressiveness. You may be dealing with a situation where the use of force or even assertiveness is useless. We all know people who are unreasonable and belligerent. Some of us are even related to a few of these individuals. Angry responses may not be appropriate in present circumstances. In delicate situations, humor or evasiveness can be the best way to handle difficult issues. Do not allow yourself to become a victim of your own anger or the recipient of another's negative energy. With psychological insight and understanding, you can become immune to negativity. What is understandable becomes less threatening. Within this scenario, the lack of response becomes a conscious choice.

You will probably tire more easily this year. There are two possible reasons for this. If you are continually involved in frustrating situations, your energy will be drained by the conflicts. But if you are absorbed in a project of your own choosing, you may need more quiet time to contemplate your next move. If you are actively pursuing your dreams, creating your vision, you are likely to have plenty of energy.

CHAPTER SEVEN

JUPITER IN THE SOLAR RETURN CHART

Introduction

Jupiter moves around the solar return chart from year to year in a clockwise fashion. It tends to move two houses each year, but positions are subject to variations depending on Jupiter's speed, its zodiacal distance from the Sun and signs of long and short ascension. The overall pattern of movement is three houses clockwise (keeping up with the movement of the Sun), and then one house counterclockwise, consistent with Jupiter's forward movement through the zodiac in one year's time. As a rule, Jupiter will be in a new sign every year because its cycle is twelve years long, the same as the number of zodiacal signs. During the twelve-year cycle, Jupiter will be retrograde four or five years and direct in motion the other seven or eight years. Little meaning is attached to retrogradation for this planet in solar returns.

Jupiter has a number of possible interpretations in the solar return chart and any or all of the following meanings can manifest according to house position. First and foremost, Jupiter tends to imply a benefic event associated with its placement. How these benefits come about and the area of life affected are usually shown by the house position and possibly aspects. For example, Jupiter in the 1st solar return house shows a need to foster benefits for yourself.

opportunities will not come of their own accord. You must create your own opportunities where none previously existed. If Jupiter is trine the Sun in the 10th house, the benefits you create may specifically affect professional endeavors, or someone you work for may encourage your push for personal growth and achievement. In this sense, the originating impulse for the opportunity is shown by Jupiter's house placement, while the supporting structure or stumbling blocks encountered along the way tend to be reflected in the aspects to the planet itself. Individuals are encouraged to actively foster opportunities according to Jupiter's position.

All opportunities will probably involve expansion into new areas of activity. The primary function of the Jupiter principle is to expand beyond the scope of previous experience. One is not to remain in a static condition; growth is not only implied, but expected. Jupiter's slow and consistent movement through the zodiac implies the steady growth needed to advance from one astrological principle to another. The movement of the planet through the houses highlights different areas in which this growth might occur. Together, the change in sign and the change in house provide a kaleidoscope of combinations, mixing signs with houses so that our consciousness becomes accustomed to growth occurring in a variety of ways and in a variety of circumstances.

One of the most common activities associated with growth and expansion is the pursuit of an education. The impetus to learn is closely associated with any push to grow and our definition of learning in this case is not limited to a structured environment (school), but can include self-teaching and life-experience training. Although education is not always a consideration for every Jupiter house placement, it can be considered to have an impact on any of the houses.

Areas of expansion can easily become areas of excess and this is a problem with Jupiter. Growth can be uncontrolled and purposeless, dwarfing any push for real attainment or benefit. For example, overeating may supplant a desire for nutritious food when Jupiter is in the 6th house. The weight gain takes the place of education. Already existing situations associated with any house placement of Jupiter may boil over if extreme limitations were previously the norm. This is especially true of the water houses, where emotions may overwhelm the individual after a long period of emotional suppression. Jupiter in the 12th, of and by itself, is generally one of the more

overwhelming house positions regardless of past experience. An
Jupiter placement can signal a false god. One or more activiti
associated with Jupiter's house may be emphasized to such a degre
that a distortion of perception results, creating the false god. Fo
example, with Jupiter in the 2nd house, money and its acquisition ca
become a driving focus of attention, even for those already possessin
material wealth.

Perhaps excesses are directly and proportionately related
suppressed urges. We now zoom ahead where we previously lagge
behind, making up for lost time. The overwhelming impetus enable
us to grow and expand at a rapid pace. The task is to avoid restrictin
this growth, while staying in control. The best option is one o
channeled and structured enthusiasm.

A question of ethics and morals is often associated with the hou
placement of Jupiter. Unlike Saturn, which is more representative o
societal structures and expectations, Jupiter implies the need for a
philosophical (or sometimes religious) consistency between wh
you are doing physically and what you believe you should be doin
ethically or morally. The underlying belief is always at the center o
any conflict, not the external expectation. Jupiter by house indicat
a desire to review those beliefs commanding a strong influence ov
one's philosophy of life and code of behavior. Current beliefs may b
limiting future growth or freedom. Within this perspective, hypocri
sy and moral dilemmas are more commonly a problem than th
frustration associated with Saturnian external laws.

Planets Aspecting Jupiter

For Jupiter's aspects to the personal planets, refer to the earli
chapters in this book. For example, Jupiter in aspect to Venus woul
be given in Chapter 5, Venus in the Solar Return Chart.

Jupiter in the Solar Return Houses

JUPITER IN THE 1ST HOUSE

Jupiter in the 1st house implies that personal needs and attribut
are becoming increasingly important and future growth may depen
on the recognition of these qualities. The potential for persona
success and progress during this year is great, but only for tho
individuals who are convinced that they are worthy of the time an

effort necessary for self-development. First evaluate your assets. You cannot begin to fully implement or promote assets you do not recognize and appreciate. Self-appreciation fosters self-confidence, which is crucial to the process of initiating and developing opportunities.

Secondly, an understanding of personal needs is important since you will not fulfill needs you ignore or consider inconsequential. The urge towards self-development arises from an awareness of what you have to offer others and what is needed in return. Self-interest is the cornerstone of achievement during the year. You must believe in yourself, your abilities and your future in order to tap into the opportunities associated with this placement. Those who hang back for fear of appearing self-centered or egotistical will miss the chance to become fully motivated towards success. Now is the time to expand your horizons and test your capabilities by tackling new and more difficult projects. Believe in yourself enough to allow new experiences to stretch the limits of your expertise. Optimism, when not extreme or unwarranted, pays off.

The most important task associated with this placement is the creation of opportunities. During the year, opportunities and benefits must be self-made; they will not fall in your lap. You must actively cultivate the situation or job you want in order to attain it. This is not the year to expect a lot of outside help. You must rely on your own wits to foster an opportunity for yourself which would not have existed if not for your efforts. Obviously you will not do this without some measure of self-confidence. This is why the reassessment of assets and needs is so crucial. For example, one individual created an exercise program in her home for neighbors and friends. She needed to shape up and she was able to draw upon her background (asset) to devise a program for herself and others. Many benefitted from her enthusiasm and push for self-improvement. None of this would have happened without her motivation.

Freedom is an issue with this placement. You cannot remain confined if you are to grow beyond personal boundaries. Personality expansion depends on freedom of movement. If you have been tied down in the past, this can be the year when freedom comes automatically. For example, your youngest child may enter school, freeing your days for new pursuits. But if Jupiter is heavily aspected, you may have to push for greater freedom since your attention is divided among several different areas. Selfishness, real or imagined, may be an issue.

You can let this time slip by. It is only through your own efforts that opportunities will materialize and success will be realized. No one will push you to achieve. But if you do pass, you are likely to miss a chance for personal growth and reward, since the tendency is for things to turn out much better than expected. If there is a spiritual emphasis associated with this placement, it lies in the implication that self-help leads to Universal assistance. Trusting in your own abilities while allowing a sense of spiritual purpose to guide you can make things happen. This is a time when beliefs about yourself and higher forces in the spiritual realm can be easily implemented in practical everyday situations. Call it a spiritual tail wind. Don't just believe; act on those beliefs and behave in a manner that is consistent with your spiritual purpose.

JUPITER IN THE 2ND HOUSE

Jupiter in the 2nd house of the solar return usually indicates a good financial situation, or at least improving financial conditions. As a rule, the money comes to you through your own efforts and is probably related to your professional endeavors or increased earning power. If you are not presently working, this can be the year you secure employment. If you are already employed, you can earn more money this year through salary increases or second jobs. Those who work on a commission or incentive basis have the opportunity for a significant increase in income. For those who work in sales, this can be the year you land that big account. The possibility of receiving a lump sum of money exists during the year. Generally, the money is not a gift but remuneration associated with your job. Gifts are more likely to be seen with Jupiter or Venus in the 8th house of the solar return chart. Common examples associated with lump sums of money include selling a business (if you own one), retiring and collecting pension funds, receiving an advance for a book or invention, or winning a sabbatical or grant for educational or research purposes.

Unfortunately, spending can also increase temporarily and it is common to make one large purchase or have increased expenses during the year. Sometimes you must spend money to make more money. If you wish to expand your business or your number of clients, you might buy extra advertising time or space. If you are selling your home for a profit, you might make improvements which

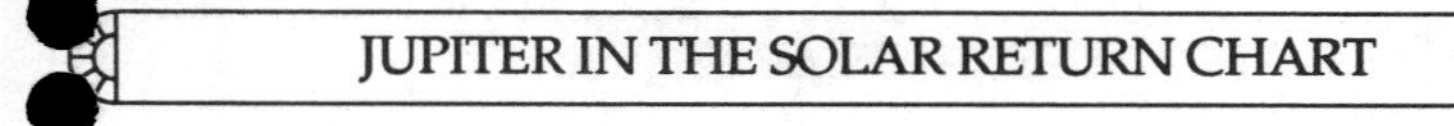

will maximize the return on your investment. The emphasis is on money-making ideas and during this year it can be fairly easy to make more money. The tendency is to be in the right place at the right time, but unfortunately, you may have to spend money to attain this position. If there are aspects from Jupiter into the 8th house, you may overspend your income and incur a large debt. The bills can reach a significant level for those who have a history of financial problems. If this becomes your situation, the push to increase your income will be even stronger.

Regardless of your line of work, the inner quality of fulfillment is stressed this year. You will not do just anything for money. Principles are important, and unless 2nd house planets are strongly aspected, the tendency is to make money doing exactly what is most fulfilling. Money comes easily when you believe in what you do well. Ethics are generally not compromised, though individual differences abide.

The task for this placement is to attract situations which give you greater self-confidence and increased self-worth. For some, earning more money can be a confirmation of one's value. For others, being successful, ethical or fulfilled becomes the confirmation.

JUPITER IN THE 3RD HOUSE

Jupiter in the 3rd house shows a great deal of daily activity. The pace of life picks up as you run around from one task to another. You have too many things to do and if the juggling act gets out of hand, you function in a state of crisis management for part of the year. It is common to have one specific project, thought or task dominate your attention and time. This interest crowds your days with activity; consequently, overscheduling is a problem. For example, a pregnant student needed to complete all her coursework before the baby was due. Papers, tests and homework took up all of her time and she had few outside interests until she was finished. Another woman volunteered to run a community event which was her whole focus of attention until the project was completed. An artist completely renovated his studio. The goal with this placement is to learn to focus only on the major tasks at hand while prioritizing lesser needs or delegating responsibilities to others. You cannot handle everything, and if you try, the day-to-day details of practical living will be easily lost in the flurry of activity. Concentration is necessary and when you are fully involved, everyday details get in your way. One tends to see only the big picture.

Despite all the activity, this is a great time to expand your mind either through school (teaching or learning), reading or writing. Many go to school or take a course sometime during the year, but a strong emphasis on being self-taught is also evident. New teachers will spend much time developing lesson plans and lectures. Writing with the hope of future publication is also possible, but the main emphasis is usually on community newsletters, magazine articles or children's stories.

Contrary to what one might expect with Jupiter in the 3rd house, this can be a mentally stressful year. Not only is the pace of life quick, the information you need to remember is greatly increased. You can get mentally overwhelmed by all the things you must do and recall. To further compound the stress you are already feeling, you could be involved in situations representative of ethical-moral dilemmas. If so, these dilemmas fall into the gray area of decision-making and are not simple black-or-white situations. Decisions involve major commitments to lifestyle patterns, and choices may be limited. You are forced to make the best decision possible given the present circumstances. Common issues include dilemmas associated with sexuality, questions of loyalty, spiritual applications in daily life, abortion, and conscientious objection. Ethical-moral dilemmas intrude on your consciousness, causing further stress in an already hectic life. Eventually the mind can feel overloaded by too much mental processing.

Learning to apply spiritual concepts to daily life situations is an important task for the coming year. It is not enough to believe in a philosophical concept; the concept must have a practical, mundane application to your life situation. Concepts may be tested by an ethical-moral dilemma. Only successful applications of spiritual concepts will lead to fulfillment and confirmation of your beliefs. Impractical ideals will fall short of satisfactory implementation. These pseudo-philosophical principles not only fail to improve one's circumstances, they also create additional stress when one unsuccessfully attempts what is unrealistic. Consistency above and below is the goal. You cannot be hypocritical this year. You must practice what you preach, and if you cannot, then you must preach what you practice.

JUPITER IN THE 4TH HOUSE

Jupiter in the 4th house indicates the need for increased peace and security in your home. It is important to feel rooted and protected physically, emotionally and spiritually by the home environment. Many individuals choose to begin a beautification program to make their living quarters more comfortable. Although projects may be expensive, there is usually a minimum of disruption. Improvements are not necessitated by a need to make repairs (more likely with Saturn), or the need to make major changes in the structure of the house (more likely with Pluto). Projects tend to be minor, and generally the domestic situation is not disrupted by the improvements. Renovations or additions are not as likely as changing the decor with paint, wallpaper and new furniture. The object here is to enhance the surroundings both inside and outside of the home, not to redo the entire house from scratch. Outside improvements include landscaping the grounds or creating a small garden for meditation. Individuals having this placement wish to create an environment which not only suits the physical needs and comforts the soul, but pleases the eye.

Security is also important; depending on the individual's age, issues can relate to present or future needs. Domestic security is the immediate issue, and you may be striving to purchase a home or secure your position in the present one. Heavy aspects to Jupiter in the 4th can indicate other influences which might support or threaten your homelife. For example, after one man's parents died, he inherited the family home and for the first time owned a piece of property. As he settled into his new home and began to redecorate the interior, his future bride informed him that she did not care for the house and wished to live elsewhere (aspect from the 7th house). This man felt the need to balance his commitment to his fiancee with his desire for a sense of rootedness. The home had become very important to him, since for the first time in his life he was working steadily towards completing a major project without abandoning his goal prematurely (aspect from the 1st house). He needed to work during the year to balance his needs and the needs of his fiancee with the desire for rootedness. Married individuals with this placement who are in the process of separating need to find and secure a home for the future. If Jupiter is very stressed, this may be a complicated task.

For older individuals, issues concerning retirement and pension plans are important. The emphasis is on IRAs, mutual funds and retirement packages, but usually security issues do not apply to funds alone since included also is a concern for the home. You may pay off your mortgage, refinance your loan at a lower rate or purchase an insurance policy which will cover your payments should you become disabled or unemployed. This is the time to make changes in your home and in your financial planning which will insure the stability of your domestic situation both now and in the future.

Occasionally Jupiter in the 4th house indicates an increase in the number of people living in your home during the year. You expand your home by taking in another person, either a boarder, roommate, foreign exchange student or relative. Generally, the addition does not indicate an increase in responsibility. The other person is able to meet his or her own needs and contribute to household chores and expenses. It is likely you will enjoy his or her presence although it can cost you extra money to have this person around.

Domestic responsibilities tend to lessen as individuals gain independence from other family members. Children may enter school or go away to college. If you were saddled with a lot of responsibility during the previous year because either of your parents or some other family member was ill, it is likely that his or her health or situation will improve at this time and your duties will decrease. Parents who are very elderly could be placed in a nursing home, or outside assistance may be brought in to help. In cases where the death of a disabled person is imminent and could be considered a blessing, this is the reason why responsibilities lessen at this time. For one reason or another, the difficulty or demand for extra attention is no longer evident.

Feelings run high and emotional swings are likely. Excessive responses are the norm and both highs and lows are evident. You could seek to suppress emotional responses, fearing the strength of your feelings and the loss of control. Many times there is a situation during the year which heightens the emotional nature and forces you to deal with the adequacy or inadequacy of your responses and those of others. If you are not in a fulfilling lifestyle, emotions will tend to be negative as well as overwhelming. But if you are actively working towards a fulfilling lifestyle, emotions can be very pleasurable and the support you receive from others is reassuring.

JUPITER IN THE 5TH HOUSE

The overall interpretation of Jupiter in the 5th house is expanded self-expression. There are a number of ways you might seek to manifest this growth. Most people begin by letting their inner personalities show more in daily-life situations, but then go on to branch out into new fields of interest. For example, one woman took flying lessons and learned to love flying. She subsequently took a job selling airplanes at a small local airport. She left a job in a family-run business which had always stripped her of time and energy. For her, flying became the symbol of her own breaking away. Being free to express yourself naturally is an issue with this placement, and you will tend to seek out those situations which allow for greater latitude and fewer restrictions. Then, as you grow to express yourself more and more during the year, you not only gain self-understanding, but you also learn to accept who you are as a person.

Artistic endeavors and creative or inventive pursuits can help to foster the externalization of inner traits and the expansion of personal expression. This is the time to experiment with new techniques and media. The freedom and risks associated with a new creative endeavor parallel the freedom and risks in an evolving and expanding personal expression.

The possibility of a love relationship exists with Jupiter in the 5th house and this is another way individuals seek greater freedom. You may be attracted to those who are already free themselves and can therefore help you to become freer to grow into new areas. Sometimes the focus of growth is not in a specific field of interest, but a pervasive desire to be free of inhibitions. Youth can be a factor in the attraction, though not always. Relationships may or may not be sexual in nature, but ethical-moral or spiritual concerns commonly affect your involvement, and monogamy or loyalty to one person might be the issue. Regardless of sexual involvement, the moral right to free self-expression takes precedence. No one has the right to control and stifle another's personal or artistic creativity, and those individuals caught in a restrictive partnership are more likely to look for a way out and somebody to assist in their escape. Spiritual connections between lovers could enhance the attraction and love, regardless of whether or not both have an interest in the spiritual realm. Relationships sometimes function on this level with one or both lovers intuitively understanding the other, and one or both generating philosophical insights and concepts leading to a spiritual connection.

Children also require greater freedom and this can be the year they go off to daycare, school, camp or college. This change affords parents more free time as well. If you have experienced difficulties relating to your children in the past, conditions may improve this year and it is a good time to open up the lines of communication. Jupiter aspected by a planet in the 2nd house implies large expenses asso ated with educating or supporting your children, and one might wi to save for this expenditure if it is anticipated. Your children have some very positive opportunities and they should be encouraged apply for scholarships and awards, or to place themselves in po tions for advancement or recognition. This interpretation is contingent on their age, qualifications and field of expertise.

JUPITER IN THE 6TH HOUSE

Generally, your job tends to get both easier and more enjoyab while Jupiter is in the 6th house. How this comes about varie Working conditions can improve and good co-worker relationships are possible. Friendly interchanges with clients, customers, fello employees, and the cultivation of a helpful attitude could prove ve beneficial to you personally and the general climate of the office. During the year, physical as well as attitudinal changes specific to t environment can enhance your surroundings, making them mo pleasurable. For instance, your office might become smoke-free, you might move to a bigger or nicer location, or new procedures cou streamline your workload.

You function at your best when you are given the freedom to handle tasks in your own way and at your own speed. Flexib schedules might be instituted, allowing you to set your own hou This flexibility may arise because of your position, seniority, or changes in office policy. You may want to take time off or cut back hours, which should not be a problem unless Jupiter is heavi aspected, implying a complicated situation. With this placement, the workload is often shared with co-workers who cover for you wh you are gone. It is also possible that the job situation is such that y can easily take time off. You might need to do this to handle other pressing responsibilities.

Job benefits, incentives or awards are possible with this pla ment, and you can benefit directly or indirectly from your position as either boss or employee. Rewards can be monetary, as in a promoti or raise, or otherwise (e.g., travel). Opportunities arise for advan ment or on-the-job training. Watch for these and take advantage of them.

Aspects from the 6th house to the 10th house can indicate that t daily running of the office either competes with or contributes to career goals. The problem here is that you will not be pushed complete long-term projects and daily short-term projects will interfere. Motivation is a problem when a laissez-faire attitude exists.

Health improves as long as you do not overindulge or make excessive demands on your body. The possibility of your health improving is most noticeable immediately following a difficult year punctuated by health problems. New or old difficulties arising during the year are generally directly associated with present or past overindulgences. Excesses of all kinds are a danger, even sudden and excessive exercising since you may push yourself to the point of injury. Included also are alcoholic and dietary excesses. Because of the tendency towards overeating, you can gain weight, but it is just as likely, and perhaps even more so initially, to try to eat a nutritious well-balanced diet. This is the time to pay more attention to what you are eating. Becoming mindful and better informed will make you an educated consumer. Since Jupiter rules higher education, you may be less likely to eat junk food and more likely to eat foods that are good for you.

Unfortunately, Jupiter's negative manifestation is the inability to exist for long in a state of moderation. Jupiter in the 6th is a perfect set-up for "yo-yo" dieting. You may lose a lot of weight, only to gain it all back if you have not assimilated the knowledge which originally created the loss. It is knowledge that supports your most positive actions and establishes a sense of balance and moderation leading to good health. Creating natural limits through educated responses to stimuli is the task of this placement.

JUPITER IN THE 7TH HOUSE

The best phrase to describe Jupiter in the 7th house is, "Ask and you shall receive." Others naturally come to your aid, giving more assistance and support than you expected or even thought you needed. If you are in a difficult situation, others readily offer to help. This can occur without asking, but the possibilities become even stronger if you not only recognize and foster the help available to you, but also ask specifically for what you need. Somebody may be in the perfect position to actualize your goals, and partnerships of this nature enable you to accomplish together what you would not have attempted alone. Help may be either tangible or intangible. Under the best possible circumstances, you benefit directly or indirectly from others in more ways than one. For example, if you are married, you

benefit through your spouse since what he or she receives trickl
down to you.

Individuals offer their insight into your problems or situation
and you progress through what another knows or has learned. Th
objective view helps you to see and interpret life more clearly. Usuall
the message is encouraging. Exchanging knowledge is importan
with this placement, but this is not all that is exchanged.

Relationships can improve; good marriages become better. If yo
have had a problem relating in the past, this interpretation is particularly apparent. Difficult marital problems tend to be resolve
especially if professional help is sought (individual or marital cou
seling). Under therapeutic conditions, those you are involved wit
become more apt to listen to your complaints and make concession
Perhaps you are also more insightful and better able to explain yo
position. Great strides can be made because something suddenl
clicks in your consciousness, adjusting your pattern for intimac
Regardless of your marital situation, good relationships with a va
ety of people in a variety of situations could exist. If you are able t
create a climate of compatibility and harmony, your success in an
endeavor during the year can be augmented by cooperation wi
others. If you are unable to work with others, you will lose out on thi
experience of mutual gain.

Sometimes personal benefits are not realized because of th
overwhelming emphasis on the partner's needs to the exclusion o
your own. In negative situations, personal needs are not on
dwarfed, but drowned out completely by the interests of anoth
Relationships can be so excessive that one goes overboard in a
attempt to please. Even what initially appears to be a benefit ca
negate your self-interest and ultimately fulfill your partner's requi
ments only. Within this context, benefits cannot be considered a
good. For example, if your spouse is offered a position overseas, yo
may be thrilled about the possibility of living in a foreign country, b
you may be forced to give up a lot in order to go. You need to asses
the effect leaving has on your long-term goals. You may be able to s
aside your personal needs for one or two years' time in order
benefit from a new experience coming to you directly through th
spouse. But instead, it may be difficult, if not impossible, for you t
leave at this time.

Any opportunity can be a mixed blessing, and the lines of stres
might be denoted in the solar return chart by the aspects to Jupiter i

the 7th house. The balance is very delicate. You may benefit greatly or feel overwhelmed by circumstances that are not your own or are inconsistent with your needs for fulfillment. One must control the swing of the pendulum. In the worst-case scenario, you will not be aware your own needs are being overrun. It is at this time that someone will step forward and make an observation that will change your perspective, if you are willing to listen.

JUPITER IN THE 8TH HOUSE

The 8th house is commonly associated with the earnings of others while the 2nd house is more indicative of your own earning power. Therefore, Jupiter in the 8th primarily focuses on the money made available to you after someone else has already earned it. During the year, your financial situation may improve specifically because you are now able to share resources with another person. If you have already been sharing resources up to this point, the dollar amount of the shared funds is likely to increase. This placement commonly shows money coming to you through others, possibly in the form of a lump sum. For those who are interested in making a large purchase such as a home, combining incomes and savings makes the goal a reality.

Besides shared resources, you may acquire money through other means. These other means include an inheritance, gift or monetary settlement from an insurance claim or legal case. Funds from profit sharing, royalties, and advances for books or inventions are also more commonly seen in the 8th house, even though one would think they would be more consistent with 2nd house earnings. Because of the extra money coming in, this may be a good time to place some funds in an investment account or hire a broker to manage your money. In this way your money works for you, creating extra funds through means other than salary.

Most likely, if you share incomes with another, you must also share his or her debts, and unfortunately, this person can be a big spender. The negative interpretation for financial concerns associated with this placement is that you are just as likely to experience increased expenses and debts caused by the other person as you are to have more money available. The pendulum can swing either way. The probability of struggling over expenses and debts increases if there are oppositions running between the 2nd house placements and Jupiter in the 8th.

Ethical questions relevant to sexual morality are important and must be addressed. Situations you face this year are meant to force you to further define your own moral code. Therefore sexual opportunities you encounter, regardless of whether or not you take advantage of the experience, must be understood within the context of your own religious or philosophical ideals. Issues concerning sexual freedom, preference and practices are most noteworthy. Regardless of your age, the cycle is such that either you must look at these concepts for the first time or you must review what now can be considered preconceived notions. For example, a young teenager was faced with the possibility of her first sexual encounter, while a much older woman who had always defined her relationships as "open" now began to desire a monogamous relationship with her lover. Each of these people dealt with the philosophical aspect of the dilemma as well as the actual feelings, thoughts and events.

For those who have the training or innate ability to make the leap to psychological understanding, the sexual and/or monetary situations and issues encountered during the year become a springboard to greater knowledge of the inner motivations and unconscious complexes of others. The intensity of relationships in these two areas of concern fosters great insight when one is able to read the subtleties of the mind.

JUPITER IN THE 9TH HOUSE

There is a strong emphasis on spirituality with this Jupiter placement. It seems that an individual naturally begins to look for a higher understanding of the life purpose. Study of a spiritual nature is frequently associated with Jupiter in the 9th, but teaching and direct application of principles are more likely for those who are further advanced. The understanding of the word spirituality should not be limited to religion and philosophy, but should also include all New Age, esoteric, and metaphysical studies. In fact, this may be the more common emphasis. It is time for the great awakening. One does not merely learn with this placement, one begins to know and experience the force and draw of the higher realms. To do this, you can be a born-again Christian or a daily meditator. It makes no difference. It is not the teachings that are most central to the growth in consciousness, but the experiential process which fuels a desire to know and understand. The pull can be very powerful for those who are in a space and time to readily open up and go with the experience.

Aside from the spiritual emphasis, educational opportunities can exist. Depending on your level of expertise and search for knowledge, you can either teach or learn. Those who are interested in learning are just as likely to be self-taught as to attend a school. In fact, the greater emphasis appears to be on individual courses taken separately, (e.g., tutoring, independent study, or self-help groups whose focus is on personal advancement through shared-learning experiences).

Although one would expect a formal educational environment, attending college is a manifestation more commonly seen with Saturn in the 9th house. The degree of freedom associated with Jupiter is more likely to lend itself to a relaxed and enjoyable study environment without the detailed structure of a degree program; however, the possibility of a formal education does exist. If you do return to school, you may receive some educational honors or awards during the year. It is to your advantage to seek recognition for educational endeavors by entering writing contests, art shows, etc., according to your area of interest.

Your mind is very active during this period and needs stimulation; consequently, there is an interest in a variety of topics. Sometime during the year, you may be overwhelmed by the information you are trying to master or understand, especially if the experiential side of the manifestation is strong. If you choose not to study this year and have no other mental outlets for your mind (reading, writing, contemplation), this can be a time of mental restlessness. One way or another, you need to satisfy the mental hunger.

In conjunction with the emphasis on teaching is the possibility of writing a book for publication. Books, especially informational, nonfiction works, seek to allow others the chance for independent study of a topic. Although publication of the book will probably not occur at this time unless there are strong placements in the 10th house consistent with this interpretation, this is a good time to begin investigation into the publishing field.

You might travel during the year, but the focus of attention seems to be more on the above-mentioned areas of concern; however, interest in a specific culture or ethnic group can be evident without any travel whatsoever. You may be fascinated by foreign customs or foods. If you do not travel, you can still follow your interests in this area through books, foreign films, or ethnic sojourns. Relating directly with a person of foreign nationality is the most likely trigger for this interest.

JUPITER IN THE 10TH HOUSE

Jupiter in the 10th house implies an easy flow of opportunity and advancement in the career arena. The harder the person has worked for this success in the past, the greater the opportunity for advancement now. Those who have done little to foster their own success might still advance, but if so, they are more likely to be at the beginning of a professional climb and will not be able to go as far as those who are better prepared. Regardless of your position, a chance for advancement can materialize without a lot of work on your part. Opportunities may specifically relate to a new field of interest, and moving into a new career arena can be so easy that one only realizes the transition in hindsight. This is a time when most career maneuvers can be accomplished easily. Sometimes little or no effort is needed to acquire a new job. You can be at the right place at the right time. You can also be released from a difficult employment situation of the past.

If you are self-employed or in a position where the development of a clientele is important, you will most likely find that, by the end of the year, you no longer need to search out clients; they come to you of their own accord. If you own a business, you may expand your operations into new areas, open a new branch office, relocate the factory, or seek a more beneficial working environment. Both owners and employees can use professional connections to network with others in the profession.

Those intent on staying with a chosen profession or job can enhance their present situation by seeking in-office opportunities for advancement or options to improve daily procedures. If you have performed well in the past, you may expect to be promoted. Transfers and travels to new locations, even overseas, are possible if they are self-initiated and welcomed. Past or future educational goals can play a role in governing your options. Educational benefits or on-the-job training opportunities could be made available to you. In the best possible scenario, the career becomes a point of stability and success in the life pattern for the year. If you are unhappy with your present position, this is an excellent time to change jobs or careers.

Too much of a good thing is the negative manifestation with this placement. Short-term goals may overrun long-term career needs. You may be so busy advancing that you skip over necessary steps or compromise your own business ethics. Advancement may require that you maintain jovial relationships with authority figures, and this

may or may not be in your best interest in the long run. Analyzing the lines of stress coming out of the 10th house can help you to envision the possibility of being put in a compromising position and the implications of such a dilemma. Opportunities abound and one does not necessarily have to settle for less.

JUPITER IN THE 11TH HOUSE

The number one issue with Jupiter in the 11th house is the issue of freedom as it relates to one or more specific areas of life. Goals are being set early in the year, and most likely the goals are meant to directly benefit the individual. The goals may include educational pursuits which will enhance career potential, but this is not always the case. In order to fulfill the goals, the individual must fight for freedom in one or more specific areas of life. The specific areas are denoted by the houses with Sagittarius and Pisces on the cusps. The symbolism of either one or both houses will work. For example, one young man wanted to attend graduate school but his work demanded that he put in much overtime. He needed to fight for freedom from the grueling hours before he could attain the educational goal he wished for. Sagittarius ruled the 6th house and Pisces ruled the 10th. He eventually quit his job and found part-time employment.

A young woman also wanted to go to school but needed money from relatives. Deep underlying family complexes made it very difficult for her to address the issues clearly and acquire the much-needed financing. Sagittarius ruled the 8th and 9th houses, while Pisces ruled the 12th. She eventually applied for scholarships and loans and entered the school of her choice. The issue of freedom is crucial to success. If one cannot master the tasks necessary for eliminating restrictions, one will not be free to accomplish the goal. Those who are frustrated may function erratically. Freedom in this case is a negative reaction to a restrictive condition, rather than a quest for wholeness.

Friends may be crucial to the goals you set during the year. They can directly assist you or help you to make the necessary connections. It might be more important who you know than what you do, and therefore networking is to your advantage. It's possible that goals can only be accomplished through a combination of energies coming together from several different individuals. The pooling of resources enables all to succeed as a group while maintaining some measure of individual success. Group efforts and self-help programs are consistent with this placement.

JUPITER IN THE 12TH HOUSE

Jupiter in the 12th house indicates the possibility of an overwhelming influence. Many times it is the emotional nature which appears to overwhelm the individual, especially if he or she is already dealing with a difficult situation, but any area of life can be difficult to control. The natal house which falls in the 12th house of the solar return chart may give you a clue. It sometimes occurs that one activity, person or theme overshadows all other self-expression and sets the tone for the year. The 12th house rules the unconscious, and perhaps Jupiter, the planet of growth, in this placement implies that the unconscious nature can grow out of proportion to the rest of the psyche. In some instances, this might be a beneficial development. Those who compose melodies might need to unlock a feeling to create the tune. Those who work with sick children need to function with a high level of compassion in an emotionally charged environment. But in other situations, psychological stress can cause a great imbalance in the psyche or lifestyle. Individuals in these situations might feel immobilized. Some are not able to function without the support of a therapist, and difficult emotional problems and anxieties can get worse during this year. Strong feelings drown out reason, especially if Mercury is conflicted in the solar return chart.

Jupiter in the 12th is like an "ace in the hole" and those having this placement may actually have it better than is readily apparent. Despite what is implied, things cannot be as bad as they say. For some the ace is a belief in God as a universal protector. Optimism and a divine sense of protection cloak all fears and give great encouragement. Jupiter points to the development of real faith, an empathic connection to the Oneness of life, and a confident trust in the Ultimate Good. Inner states associated with meditation, spirituality or religion may be particularly comforting. But keep in mind that both the use and abuse of spirituality can be seen with this placement. One is able to twist ethical situations and misuse spiritual concepts for the sake of personal gain. This is all done quietly, behind closed doors, and the truth might never be known. The end can justify the means and in this case the ace in the hole becomes the last laugh.

The day that Jupiter crosses the Ascendant by transit is especially important since it indicates a time when those activities and processes kept behind the scenes can become a reality. Problems can intensify as Jupiter crosses into the 1st house, but generally this is a rewarding time and you might wish to schedule beneficial events around this transit.

CHAPTER EIGHT

SATURN IN THE SOLAR RETURN CHART

Introduction

For those individuals who remain in the same location, Saturn moves clockwise around the solar return chart two to three houses each consecutive year. Frequently it moves consistently from one quadrant to the next, stressing responsibility in different kinds of social interchange. Movements are somewhat modified by signs of long and short ascension and Saturn's direct or retrograde motion.

Saturn rules reality and the reality of any given situation is that we are all ultimately responsible for creating our own lives. The responses and choices we make to any given situation, together with the fears we avoid, form the backbone for our pattern of living. Only when we accept the responsibility for creating our own fate can we be freed from the necessity of living it. Only when we face our fears can we live unafraid. Saturn is the key to this change in orientation. It is the ultimate reality, the ultimate fear, the karmic avenger, and ruler of the universal laws of nature. It is through Saturn's house placement that we come face to face with ourselves, the structures and limitations of reality, and the laws we must live by.

Growth is a complex process, occurring in a variety of ways. Saturnian growth begins by recognizing the existing structures we have incorporated into our lives, and possibly, the frustration or limitation caused by this edifice. It seems strange to talk about growth in terms of limitation, but astrologically Saturn rules structures which are at one point protective and at other times restrictive. Like the lobster or crab which sheds an outgrown, protective, hard shell in order to grow, individuals must periodically shed old structures grown restrictive with time. As with the crustaceans, the shedding process may be slow and laborious, followed by a period of extreme vulnerability until the new shell has formed and hardened into a protective structure. We often complacently mistake familiarity for security. We assume we'll be safe as long as we stick with the same old routine and keep doing what we've been doing all along. But what was once protective and safe eventually ends up inhibiting our evolutionary cycle of growth. Sometimes, only our frustration with things as they are awakens us to the need for change.

The movement of Saturn through the solar return chart coincides with an awareness of structure. Sometimes this awareness is associated with frustration, and the accentuation of restrictions now blocking evolution. It is through this realization process and frustration that we first become aware of the need to grow beyond our present structures. Therefore, the first step in the Saturnian process towards growth is recognizing structures, frustrations and limitations imposed by our present patterns of living, particularly in those areas of life signified by Saturn's house placement in the solar return.

The second step towards growth is reassessment of the situation. A review of the facts yields further information, options and understanding. Only by defining the inhibitors to the evolutionary cycle can we move toward resolution and continued growth. Self-criticism and constructive feedback from others may be appropriate at this time.

It is during the reassessment process that the reality of our situation becomes more obvious, as well as the laws by which we are governed. To succeed, we must deal with life realistically. We cannot twist nature to our own liking, nor can we expect others to compensate for our deficiencies and fears by solving our problems. Saturn rules the naked truth devoid of magical thinking. We must live within the laws of nature to succeed and survive. If you hate your job, quit. Do not wait for an act of God to find a new job. If you want to complete a major project, work on it. It is the only way it will get done. If you want to lose weight, diet and exercise. Hard work and facing the issues head on will get you where you want to go. Wishful thinking will not.

These issues are all totally within your range of control and you are free to act upon them, but other problems may not be yours to control. You cannot make your husband stop drinking; you can only change the way you handle the situation. You cannot change hypocrisy in others; you can only reassess your own beliefs. During the reassessment process, develop realistic options which could feasibly lead to a resolution of the problem or to a change in the way you handle the problem. Successful solutions are those which are practical and realistic —that is, they conform to the laws of nature.

Once you understand the problem and the options available, you must assume responsibility for the outcome. Each man or woman contributes to his or her own fate. As the partial creator of any problem, you also have some control over the solution. But assuming responsibility for your present situation can be more threatening than blaming others. Some prefer to buck the system (defy the laws of nature) and complain, rather than take action. Remember, familiarity breeds a false sense of security and many prefer to hang on to their restrictions rather than venture into the unknown. They learn to live with their frustration rather than grow beyond it. Consciously making a decision implies assuming responsibility for your success or failure.

Facing your fears concerning success or failure is ultimately what Saturn is all about. At this point, you are very aware of the reality of your situation, and you have a list of options for the future, not all of which may work. You understand your own contribution to the situation and what you must do in order to move toward a resolution. You have seen your own shortcomings and become aware of your darker side. You comprehend the laws of nature relative to your situation and know everything is earned in this case, nothing is given, and nothing is guaranteed by the Universe. You now realize that you must stick your nose out and take a calculated risk. In order to progress, you must face your fears, and move toward your nightmare rather than back away. It is only by passing through a period of vulnerability that one can break out of a protective yet restrictive structure and progress into a more suitable environment. You give up the security of the familiar and step into the unknown.

The test of any solution is that it works. Even chronic problems need to be solved and not just cosmetically concealed. Success requires much hard work and the systematic testing of solutions until the correct combination is found. Consequently, success may not come easily. If the first solution does not work, you must go on to test another and another. Thus with each failure begins a new cycle of awareness and reassessment, ultimately leading to a well-deserved success. There is plenty of room for hard work, discipline and perseverance in the area of life indicated by Saturn's house placement. Eventually, every area of life will be touched by the need for reassessment and resolution. Chances are the time for reevaluation coincides with Saturn's placement in a particular house of the solar return. In that area of life, you must go above and beyond what is normally expected.

Planets Aspecting Saturn

For Saturn's aspects to the personal planets, refer to the earlier chapters in this book. For example, Saturn aspects to Venus would be given in Chapter 5, Venus in the Solar Return Chart.

Saturn in the Solar Return Houses

SATURN IN THE 1ST HOUSE

Saturn in the 1st house is a self-imposed limitation and it is likely you will restrict yourself or your behavior for one reason or another during the coming year. It may be that you are trying to complete a long-term task or you may take on a major responsibility. You will tend to do all of the things you have to do and none of the things you want to do. This is a good time to push yourself to finish a master's thesis, doctoral dissertation or book, or to start your own business. Any piece of work requiring long hours, discipline, organization and perseverance can be completed while Saturn is in the 1st house of the solar return chart. The project is likely to take most of the year to finish and the earliest completion date is three months before your next birthday.

You want to assess the reality of your innate and developing abilities. To do this you can concentrate on a major project. Those who are not seriously involved with the completion of a specific body of work can instead assume extra responsibility which may or may not involve other people and can go beyond what is normally expected. For example, suppose you have a newborn baby (which is demanding enough by itself), and you also decide to be a nursing mother during the year. Anyone who has ever nursed a baby or been involved with a nursing mother knows how this can tie you down and seriously limit the amount of free time available. However, you are the person who is deciding to nurse the baby and therefore you are the one who imposes this limitation on yourself. This is also the case if you are involved in a major project. You are the one who limits your outside activities and social life in order to accomplish the task at hand.

The negative manifestation associated with Saturn in the 1st house is the ability to impose limitations because of an overriding fear. You might shy away from particular opportunities possessing great growth potential simply because you are afraid to repeat a situation reminiscent of the past. Events occurring during the year may mirror former events, allowing unresolved issues to surface once more. In the negative manifestation, self-imposed limitations have nothing to do with the successful completion of a task or progress towards the future, but they have everything to do with stagnation and avoidance behavior. You can restrict your activities simply because you are afraid to face your fears, grow beyond your limitations and take risks.

If you meet and conquer your fears during the year, the rewards can be great. You can free yourself from unnecessary restrictions and gain more control over your life and destiny. But if you avoid the fear within you and limit your behavior unnecessarily, you will not only maintain your present level of fear and limitation, you will also augment your conflict for the future.

In your quest to avoid a confrontation with your fear, you will also have to avoid certain people and certain situations. Isolationism in one form or another is associated with Saturn in the 1st house. In the negative manifestation, withdrawal is used to avoid that which you fear most. In the more positive manifestation, withdrawal is necessary if you are to complete a body of work or fulfill your responsibilities to others. You are capable of being self-sufficient during the year. You are the one who must take total responsibility for your own destiny and actions. You cannot depend on others to accomplish what you alone can do. Nor can you expect others to undo what you have done to yourself.

Minor illnesses are sometimes associated with Saturn in the 1st house but usually these illnesses are minor chronic inconveniences. Some examples are perpetual sore throats, numerous colds, problems with teeth and bones, and back problems. These illnesses are annoying, but generally they are not incapacitating. They reflect stress on the body and emphasize the need to limit activities to those experiences which are truly important. It is unlikely that you can sustain your new level of responsibility and productivity while still maintaining your former commitments. It is essential that you take care of your body and reassess your level of activity. Minor illnesses are the body's way of telling you to slow down and lighten up.

If you ignore your body's needs, or refuse to direct your energy into meaningful and healthy experiences, it is possible that an illness can make it physically impossible for you to continue. This is especially true if you actively avoid certain situations and refuse to confront your own fear. Fear and tension patterns can directly coincide with patterns of illness. The stronger the fear and the avoidance behavior, the more likely you are to be ill during the year and the greater the possibility of that illness being serious. If you sensibly meet your responsibilities, realistically face your challenges, and get enough relaxation, you can maintain good health.

SATURN IN THE 2ND HOUSE

Saturn placed in the 2nd house is generally not a sign of financial success and monetary abundance. It usually means learning to live with less either because you have to or because you want to. Individuals have experienced this transit differently. Some have experienced financial hardship because they have overspent in the past, but others have planned inadequately for the future and are now faced with a pending major expense. In either case, the reality of the financial situation becomes very evident and these individuals must suddenly take fiscal responsibility for their past and future actions and curb their material appetites.

Other individuals with this placement in their solar return charts are not so hindered by budgetary limitations because they have priorities stronger than money. To them, money is seen as only a means to an end and not an end in itself. Fiscal control is necessary for goal fulfillment, and there are priorities more important than money. Those individuals considering major changes in lifestyles will have this placement if the change involves a decrease in salary. Goals of returning to school, having a child, starting a business, working on a long-term project with little financial reward, retiring, or moving to another part of the country are consistent with Saturn in this house. Its placement here signals your ability to cope with financial belt-tightening.

Use this time to reassess your value system and establish priorities consistent with your future goals and present earning power. Do not allow your options to be limited by inadequate funding. You may have to budget your income, save for a rainy day and cut your living expenses in order to realize your dream, but the adjustments can be made and you can live on less. This time can be used to save for a major expense (such as purchasing a house or car). It is also possible that this is the first year after a major purchase when finances are generally very tight.

The 2nd house is the house of self-worth, and during the year you can place yourself in a situation where you are either underpaid or not fully appreciated. If you have a poor self-image, you may need to experience deprivation to get in touch with your own real value. The fact that your situation is not healthy encourages indignation and reassessment.

Those individuals with strong feelings of self-worth might stay in a low-paying job or difficult situation for different reasons. Moral values and ethics are extremely important and you may deliberately stay in a difficult or low-paying position because of your ethical priorities. A man working for a non-profit organization stuck with his job despite the financial hardship placed on him and his family. His belief in what he was doing was so strong that he could not in good conscience leave until his task was completed one year later.

The process of reassessing your value system and living by the financial priorities you set yourself is consistent with Saturn in the 2nd house. Money and material possessions should not matter, only your priorities for the present and your goals for the future. During this year, long-term goals and satisfaction are more important than short-term compensation.

Saturn in the 2nd denotes a desire to limit all excesses and overindulgences. If you are a smoker, heavy drinker or drug addict, you will be very aware of the connection between your behavior now and the possibility of present or future health problems. The urge to limit these indulgences will be especially strong early in the year and it is to your benefit to take action at that time. (The desire to quit smoking is especially strong with this placement.) You will begin to see the negative consequences commonly associated with your behavior. These negative reactions will manifest both in your own body and in the health of others. The issue of control will surface again and again. You can go cold turkey with this placement, simply consider this option, or seek professional assistance.

SATURN IN THE 3RD HOUSE

Saturn in the 3rd house indicates a year of much study or mental work. You take your studies seriously and knowledge is probably crucial to what you hope to accomplish in the future. This is an excellent time to concentrate on difficult mental tasks or complex topics of interest. You can be involved in a major project or course requiring concentration, persistence, organization, and many hours of work. Typical commitments include taking a course, organizing community projects, formulating new ways of thinking for yourself and others, writing a book or paper, studying for qualifying examinations, or consulting with others on a regular basis in order to gather new information and insights.

You may finish your endeavor during the year, but if Saturn falls in the 1st house of the solar return chart next birthday (which is most likely to happen), your work will probably require another year's effort. (Saturn in the 1st house is commonly associated with the completion of a long-term project and is a good placement for those interested in focusing on a major achievement.) Your mind is constantly at work and it is best to choose a positive focus. If you are studying, learning or working on a major mental task, you will be less likely to feel depressed or anxious. You need a strong mental focus this year to use up excess mental energy.

Major decisions usually arise during the year when Saturn is placed in the 3rd house of the solar return chart. Responsibility to yourself and for another person is commonly a factor in the decision-making process. You may be making a decision for two and not just for yourself alone. Dilemmas are common since your needs may directly conflict with the needs of your counterpart. Reality assessment is crucial. You must realistically weigh and evaluate all the intricacies of your situation and actively search for options. Options are important, but unfortunately, looking for new possibilities is an insight most people tend to miss when dealing with tough issues and this Saturn placement. Rigidity of thought is a common flaw. Your accurate evaluation clears the way for the most successful resolution of your problem, even when there can be no clear-cut success. If you are mentally stressed by the problems you must solve or the decisions you must make, it might be in your best interest to discuss issues with someone who is impartial and not threatening. Seeing a counselor at this time can help you to do this. Your insight should be increased through the interchange.

Your inability to see positive solutions and options to your dilemma can lead to depression. Depression is sometimes associated with Saturn or Sun placements in the 3rd house of the solar return chart. (Severe depressions are more likely with the Sun.) Depression is the mind's way of calling attention to what needs to be corrected, decided and surmounted. Depression can lead to positive results when it is used as an early warning signal and learning tool. Life is a growth process and depression can indicate your inability to seek and find newer, more mature ways of handling situations. Depressed individuals need to begin to solve some of the problems they are faced with. They should do something—anything! They also need to get in touch with their feelings and learn to communicate those feelings effectively to others. Information blockages are likely and either you cannot or will not discuss things openly. Isolationism is commonly generated by the issue or problem you are facing. If you are unhappy with your present situation, it is time to face the problem and begin to make informed decisions. If you avoid issues and responsibility, your depression can grow deeper and become more obvious to others. You will seldom praise and often criticize; your speech will become caustic. Situations will seem hopeless. Your level of activity will slow down and eventually you will be perpetually tired. As you withdraw from others, effective communication will become extremely difficult, if not impossible, to maintain.

Everyday situations can become everyday hassles that get on your nerves. Common daily occurrences are more difficult to weather. It's the little things that bother you and seem more troublesome this year. What used to be simple is now more complex, and day-to-day things that need to be done take longer. For example, if you move to a foreign country and you do not speak the language very well, handling common chores such as shopping for food, cashing a check or going to the post office are laborious until you get the system down. Similar situations are associated with new babies, living in a dormitory or alone for the first time, moving to a new location and various other common life experiences.

Problems with siblings are possible during the coming year. You may have a problem relating directly to a family member, but it is more common to have a brother or sister with a problem or issue he or she needs to handle. The encouraging side of this Saturn placement and interpretation is that the problem is being confronted. Usually this is the year your brother or sister begins to handle a situation in a responsible manner. For example, if your sister is an alcoholic, and has been for years, she could decide to enter a rehabilitation program at this time. Your contribution during the year is to support her decision and help her in any way possible. You may wish to join a program which will help you to better understand her situation and your role in her successful treatment.

If your sibling refuses to confront the issues, this can be a time when you feel forced to act on his or her behalf or completely alienate yourself from the situation. Problems need not be very difficult and can be relatively minor with this Saturn placement. Productive siblings will not suddenly have major problems or behave totally out of character. The best predictor for the future is past experience. Problems which may arise will be consistent with your sibling's normal personality, and may mark a turning point in life. This is a time when your encouragement can be very supportive at a crucial passage.

SATURN IN THE 4TH HOUSE

Saturn in the 4th can be a sign of increasing commitment and responsibility within the home environment. Usually this newfound sense of responsibility involves the physical home itself, though the manifestation is not limited to the physical structure alone since emotional responsibility is also an issue. The condition of your home, apartment or dwelling is such that you need to make repairs or improvements. If you have let your house go in the past, this is the year you will feel forced to take corrective action. Chronic and long-time problem areas will suddenly become especially annoying. Repairs can be major, though not always. If you are buying a home during the year, you may be drawn to older or rundown homes requiring much hard work and restoration. Physical discomfort in the home is common, especially before or during the repair and renovation process. If you have not neglected your house, time can be spent on routine maintenance or improvements geared towards making your house more comfortable.

External changes in your living environment tend to parallel inner emotional changes. This can be a year when you feel emotionally responsible for the well-being of certain family members, regardless of how well you get along with these people. The desire to accept responsibility for the physical state of your home is usually extended to a desire to assume responsibility for the emotional and physical state of others. If you feel you can help, you are likely to do so.

You may accept a family member into your home, or you may return to your parents' home to live or visit and lend assistance. Those who are mentally incapable of making important decisions look to you and other family members for guidance, support and possibly shelter. It is common for the health of one family member (usually a parent) to deteriorate during the year. If this is the case, you can nurse this individual yourself or provide for his or her daily needs. Elderly parents may be placed in a minimal-care retirement community or nursing home. If you have a grown child in the midst of a divorce, separation or family crisis, he or she may return home with small children in tow. Your strong sense of familial responsibility compels you to take up the slack and help out where needed. You willingly give up some measure of physical and emotional comfort in the home to help those you love.

If you are a much younger individual, you may choose this time to become a parent yourself. In its most positive manifestation, Saturn in the 4th is a tendency to respond in a helpful manner to family crises, problems and issues. Unfortunately the reverse is not always true. This may not be the best time to look for assistance and help for yourself. Even if you are in a difficult position, family members may be unable or unwilling to support you at this time. The tendency is for you to lend support rather than receive. Saturn in the 4th is not normally associated with the fulfillment of dependency needs. On the contrary, it is more closely akin to their denial. Pushing your own needs may be inappropriate under the present circumstances.

Some individuals lack the strong family ties necessary for such dedication to the needs of others. For them, no personal gratification, only frustration, can come from involvement in family issues and problems. During the year, events trigger memories of the past and old feelings of hurt and disappointment resurface. If this is your situation, you may wish to protect yourself by limiting family contacts or withdrawing completely. Saturn in the 4th can show a separation from family involvement, and if you have already exhausted all your options, including therapy, this might be in your best interest. Saturn indicates that you test your relationships and eliminate or restrict those ties which are unfulfilling while strengthening those which are meaningful.

The process of reassessing relationships for their emotional reward is not limited to family involvements. All close relationships are subject to scrutiny. Usually family relationships are the ones tested the most, but any intimate relationship can be suspect. You may find it more difficult to trust others during the year, so you look for conspicuous proof of your loved one's affection and caring. If the relationship passes the intimacy test, commitment should follow, but if you are not satisfied or reassurances are not forthcoming, you can set emotional limits on those involvements which are unrewarding and nonsupporting. Saturn rules reality and this is a time for realistic appraisal of emotional situations. It no longer makes any difference what you are told or what you are led to believe. You know when you are unhappy and it makes sense to seek practical solutions to relationship problems. If you cannot work with your partner to make corrections, you will tend to live alone emotionally, if not physically.

It is common to have Saturn in the 4th when the native is trying to recover from, or deal with, an emotionally painful relationship and therefore feels the need to set emotional limits. Individuals can elect to withdraw from situations entirely or structure involvements in such a way as to protect themselves. Those who withdraw completely assume total responsibility for their own emotional well-being and refuse to be responsible for anyone else. They do not encourage nurturing attention from others and may shun all offers. Walls are built to allow a year's worth of time for healing and recovery, but in the meantime they feel lonely, withdrawn and neglected.

In the most negative manifestation, those who build walls exhibit contradictory emotions. They expect or even demand to be taken care of. They see total dependency on others as proof of another's love, but at the same time refuse to commit themselves emotionally to a relationship or accept any responsibility for the other person involved. The goal is a one-sided exchange, an improbable situation in which all their needs are fulfilled without any fear, risk, or effort. Those who are wise realize that this is a time to assess the mistakes of the past and set guidelines for future emotional involvements. Trust and mutual responsibility are needed for emotional security.

SATURN IN THE 5TH HOUSE

Saturn in the 5th house of the solar return indicates that normal self-expression is affected by present circumstances. For one reason or another, you feel very unsure of yourself and find it difficult to express who you are without fear of criticism from others. Most likely, your social milieu has changed. You may feel like a fish out of water and you no longer feel relaxed and comfortable with yourself.

This is especially true if you suspect that you are in an unfriendly environment and you are consciously trying to be inoffensive to those around you.

For example, a psychic who went to graduate school found that another psychic had attended the same school before him. The first psychic had verbally threatened various professors and alienated them to such an extent that the second student had to deal with the stigma of being a psychic. It was to his advantage to be very inoffensive and nonthreatening until he was able to establish his own personality as separate and distinct from his predecessor's. He experienced a period of self-imposed personality limitation while he worked towards greater definition of who he was. When he succeeded he was able to show practical applications for his skills.

Another example of this uneasiness with self is the experience of an older woman dating a much younger man. She did not really fit in with his friends and he could not really relate to hers. The transition from one environment to another may involve some inhibition. The natural flow associated with self-expression seems more controlled and consciously directed. There will be situations where caution is warranted and even advantageous. The ability to "fit in" where you would not normally go can be beneficial to your own growth and to those you meet.

The need for greater discipline when working creatively is commonly seen with this Saturn placement. During the year, even the creative process itself will be subject to greater control. The natural creative flow needs to be channelled in a more organized and productive manner, while the final creation needs greater refinement. Some examples of these changes might include working within a scheduled time frame or having deadlines to meet. You may have a particular piece which is popular and can be reproduced for the mass market. You probably should review the creative process and make changes according to your future needs and ambitions.

During the year, you are likely to ask for constructive criticism of your work in the hopes of getting a clearer definition of what you are trying to get across. You may find that what you create is not exactly consistent with your intention. If this is true, the creative process will now be more labored and tedious as you work to refine your creation. You become less involved with Venusian creativity since you are now more involved with Saturnian perfectionism. Major pieces of work and long-term difficult projects are also implied by Saturn in the 5th house.

In all of these instances, it can initially become more difficult t
express yourself creatively. Blockages can occur, but the goal in th
end is a better product which is a more accurate reflection of what you
were trying to get across. You are willing to make the extra effo
necessary to perfect the creative process and, ultimately, your cr
ation. Those who are unable to make this kind of commitment and
who lack the perseverance necessary will remain blocked througho
most of the year.

Self-criticism can be at an all-time high during this period while
self-confidence may be shaken. These personality tendencies directly
relate to the refinement process and the desire to handle very diffe
ent or difficult situations. Moving closer to perfection means first
seeing what is less than perfect. Do not let criticism, whether yours o
someone else's, discourage you. You should actually welcome co
structive criticism since it will help you see what must be done. You
must go beyond your disappointment. The quality required fo
future tasks is much higher than that required in the past. You nee
to be more organized, disciplined, consistent and practical in the
future. Sloppy habits, laziness, inconsistency and hypocrisy can lim
your success and weaken your self-confidence, both this year and i
years to come.

Relationships, especially love affairs (but generally not includin
marital relationships), can be more difficult with Saturn in the 5
house. Saturn is consistent with limitations of one form or another
These limitations may be specific to this particular relationship o
they may be residual fears from a previous involvement. Commo
specific limitations include long-distance relationships, May-Decem
ber romances, extramarital affairs, or relationships that involve a
great deal of separation at least for the present year. You or your lov
one may not have the available time necessary to carry on a wil
romance. Practicality may be important.

The very worst manifestation is that you will be denied th
relationship you want. This is most likely to be true if you a
presently involved with a person who has previous commitments.
Sometimes the relationship itself lacks affection and true caring. Wi
very negative manifestations, Saturn may indicate sado-masochisti
attractions or relationships that involve more pain than love.

If you have children, they may require closer supervision duri
the coming year. More of your time and energy can be taken up b
their demands, and some individuals will find this frustrating and

restricting. Situations will require you to be more actively involved with your children on a daily basis. New parents tend to have this placement (or a similar strong Saturn placement) because of the time and energy needed to care for an infant. The most common issues associated with school-aged children are either academic or behavioral goals or problems. For one reason or another, children may not be able to cope with everyday circumstances on their own and now need your guidance and support. This is especially true for children going through major life stages (infancy, terrible twos, puberty) or children trying to adjust to major life changes (relocation, new school, parents' separation or divorce, birth of a sibling). Increased discipline may or may not be appropriate, depending on the circumstances. Increased support and guidance is always appropriate. If your children are now adults, it is still possible for them to be involved in life situations that concern you or require assistance.

SATURN IN THE 6TH HOUSE

Saturn in the 6th house can indicate a health strain. However, depending upon your age and general body condition, you may or may not experience a health problem. Reassessing your health and work habits is a positive way to use this placement and may be crucial to continuing good health. Minor health problems such as colds and sore throats are likely if you overtax your body, but as a rule, most health problems will arise from long-term abuse, chronic problems, severe dietary deficiencies, or work-related stress. Immediate changes can lead to improvements. For example, a long-time cigarette smoker stopped smoking after she experienced trouble breathing. She suddenly realized how precarious her health had become from the continuous abuse. Saturn rules reality and the reality of her situation became very clear to her. She needed to stop smoking immediately or run the risk of developing a serious health problem.

If you are having chronic health problems, these problems may be aggravated or initiated by your involvement in certain circumstances. A nurse with chronic back problems realized that lifting sick patients aggravated her condition, and as she got older, she could no longer afford to jeopardize her health for the sake of a paycheck. Other careers or even other assignments within the nursing profession could be just as profitable.

Severe deficiencies are associated with poor nutritional habits. Many people with this Saturn placement go on very strict diets which are not nutritionally sound. The tendency is to eliminate certain foods or meals from the menu entirely rather than cut down on intake. The goal is to lose weight (and this is commonly done during the year), but poor eating habits are often established in the process. The diet one tends to choose, by its very nature, seeks to distort your eating habits rather than focus on balanced nutrition. Examples of these kinds of diets are those which concentrate on eating one particular food or food group. It is true that you can lose weight by not eating certain foods, but you would be better off cutting out unnecessary food substances such as caffeine, sugar, fats, and red meats rather than those food substances which make good nutritional sense. Artificial restrictions which do not suit your body's needs can tax your health.

The best way to develop a personal diet is to become aware of your body's health from the inside out. Watch what you eat and how it makes you feel. Eat only those foods which make you feel good. Develop an exercise program that makes you feel more alive without taxing the system. Saturn represents a teacher and a taskmaster. You can teach yourself good health habits by realistically observing what makes you feel good and what makes you feel bad and making adjustments accordingly.

Stress-related health problems can also be traced to working conditions and co-worker relationships. Most commonly, the individual in poor health who does not fit into any of the three categories mentioned above hates his or her present job and/or is under too much pressure to perform. Responsibilities may be overwhelming and schedules too taxing. The job description itself may include tasks which are difficult, if not impossible, for the employee to accomplish without creating a health strain. Personal feelings about fellow co-workers, partners and bosses can also cause nervousness and/or physical stress. (For example, office love affairs and employee conflicts frequently lead to stress.)

Again, Saturn placed in the 6th house of health signals the need to deal realistically with job-related problems. If you hate your job, find another. You can start from scratch in a new position, new company or new career. If you prefer to stay with your present job, institute changes that will make tasks easier to accomplish. If you are in a position of authority, delegate responsibilities to others. A good manager knows how to develop the employee potential under his/her care. Use the talents of others wisely while reducing the amount of work you must handle personally

Regardless of your level of authority, systematize and structure office procedures under your control to make daily routines easier to accomplish. Use the suggestion box to point out problem areas and possible solutions. If you feel restricted at work, realize that the restrictions are meant to make you aware of the problem areas. If you find your job tedious and boring, you may be wasting your talents and need to move on.

Promotions are likely if you have performed well in the past and can handle extra responsibility. A teenager with this placement may decide to enter the job market since he or she is now mature enough to behave in a professional manner. He or she may instead refuse to find a job at all. (Saturn can also be associated with unemployment in rare situations.) Regardless of the teenager's decision, the issue will arise. If you love your job and naturally work hard, you can become a workaholic during the year and spend many hours working overtime. Moderation is suggested. Remember, health is an issue. If you cannot regulate the amount of time you spend working, then your health will. For good health, structure your diet, habits, and work with a realistic respect for your body, its needs and limitations.

SATURN IN THE 7TH HOUSE

Saturn in the 7th house indicates that close personal relationships are defined, limited or structured in order to establish a greater degree of safety and security. For those dating and not married, usually the relationship is defined as not ready for marriage during this solar return year. Although this can be a time of great commitment for those who are already married, there is generally a lack of commitment or a need for delay with those who have not yet tied the knot.

Relationship limitations can be either natural or artificial. Natural limitations are inherent in the situation while artificial restrictions are imposed by one partner or the other. Inherent limitations include living in different areas of the country, scheduling problems, and basic socio-economic or character differences. The time you spend together may be restricted by work, distance, parents or other circumstances. Aspects to the other houses may give you some clue as to the source of these restrictions. This is not the year of the free-flowing, easy-going relationship, since an important partnership will be reassessed and structured during this time.

The hallmark of this placement is realism. You must deal realistically with a relationship in order to develop a sense of security and safety. You must see your partner as a real person with human frailties and needs and then address those issues. Most significant relationships will have an issue which must be handled successfully if the relationship is to prosper and grow. For example, one extravagant young woman was dismayed at her boyfriend's present state of poverty. He had just graduated from law school, but had not yet become established in the world. She wanted to marry him, but did not want to be poor. (Those who are used to spending $200.00 on panty hose rarely have a sense of monetary restraint.) She began to see realistically what life would be like married to a young lawyer and she subsequently sought options and changes within herself that would resolve this issue rather than destroy the relationship. It also became obvious to her that the young lawyer would need to put in extra hours during his first few years with the law firm and the woman would need to occupy her time alone with her own pursuits until they could be together in the evenings.

The need to realistically structure a relationship so that certain tasks can be accomplished is common with this placement. The overall goal of future safety and security is a motivating force behind this process. During the AIDS crisis, a homosexual male with this placement curtailed his barhopping and also began to limit and structure his continuing involvements in the interest of safety and security.

Basic personality differences between partners seem stronger during this year, and depression or pessimism can be a problem. This is one of the more negative manifestations of Saturn in the 7th. Generally, those who are depressed are the butt of much criticism. But by far the most negative manifestation is a blatant refusal to love and share. Artificial limitations can be set by one partner or the other; usually this occurs only in the less viable relationships. The inability to build a trusting and safe relationship leads to a denial of commitment and severe restrictions; eventually the relationship may dissolve. Limitations are walls meant to insure the security of one of the individuals while compromising the partner's needs and relationship goals.

Usually the lack of commitment is a two-way street, but you may not see it that way. In the beginning, you may compensate for the lack of commitment by deliberately lowering your expectations and lessening your demands. You may limit your own behavior in an attempt to comply with your partner's restrictions, thinking that if you acted in a certain way (either by saying or not saying, doing or not doing), the relationship would survive. You accept the limitations imposed. If the relationship is basically a good one, you will outlast the difficulty. But if you continue to support an abusive partnership, you only fool yourself into thinking half a relationship is better than none. Relationships that severely restrict your expression are not in your best interest. Use this time to reassess your involvements. Recognize restrictions and evaluate their source and necessity. Ask yourself, "Is the relationship worth this, or is there another way to handle the situation?"

Saturn in the 7th house indicates your awareness of humanity and the difficulties inherent in associations with others. Relationships involve a certain amount of discipline, obligation and responsibility. Much of this will fall within the normal give-and-take of relating, as long as you do not stay in partnerships which are excessively difficult or detrimental to your psychological health.

SATURN IN THE 8TH HOUSE

The most common interpretation for Saturn in the 8th house of the solar return relates to financial concerns caused by either debts or limited shared resources. Debts are an issue, and it is not uncommon to borrow a large sum of money during the year for an expensive purchase such as a car or house. But it is more likely that you will feel pressured to pay off your debts, both existing loans and any new ones you acquire. Some individuals force themselves to limit expenditures by refusing to spend any more money until they pay off all debts. Others, who have refused to face financial reality in the past (allowing their debts to reach the credit limits), are now forced by their circumstances to limit spending. They have exhausted all their options. Regardless of whether you are working hard to finish paying off your debts or just reaching your debt peak, you will probably feel pressured to pay your bills.

Saturn can imply limited shared resources; consequently, you may receive less money from others. If you normally share finances with someone else or receive money for other reasons, this can be a year when these other forms of income (outside of your own earning power) decrease. If you are married or have a roommate, he or she may not be fully employed for part or all of the year, thereby earning less income than usual. On the other hand, he or she may be fully employed but hesitate to share funds, instead spending whatever money is available on either worthy or unworthy causes, agreed upon or not. In any case, the amount of money shared or earned by your partner can decrease for one reason or another.

For example, a young housewife with this placement was having trouble stretching her husband's paycheck to meet the bills. They lived frugally but still there was not enough to go around. He was due for a raise and they hoped everything would work out but just as he received the raise, they were forced by circumstances to buy a new car. The new car payment took up the entire raise plus ten dollars extra a month, so there was even less money to go around that year.

If you normally depend on others (possibly your parents or some other relative) for financial gifts, you may receive less money this year than you have in years past. It is probably time you grew more financially independent and became responsible for your own debts. The restricion on shared resources also includes situations associated with wills, inheritances and occasionally even bank loans. Do not expect someone to give you money unless you expect year-long delays. Saturn in the 8th is not a reflection of the amount of money you yourself earn. That is a separate matter seen through the other houses, most notably the 2nd. Your earnings may be very good this year, but with Saturn in the 8th you would still tend to have some of the above mentioned problems with the money you receive from others.

Sexual issues are possible with this placement, but generally they are rare in comparison to the emphasis on financial considerations. Sexual limitations may include abstinence, impotence or frigidity in extreme cases, or a general lack of fulfillment. Critical judgments of self or others can inhibit sexual performance. Issues outside of the bedroom, such as power, dominance, and restriction, might affect sexual interchanges. Limitations occur for a variety of reasons, but the partner's absence, your or your partner's ill health, or an inability to function well are most notable. For example, if your wife is pregnant with twins, her desire may wane, assuming you could master the logistics to begin with.

For a few, sex will be less enjoyable this year, especially if it is rushed, occurring in inconvenient locations, or under stressful circumstances. This is especially true for those who are involved in an extramarital affair. If you cannot fully relate to your partner, you may view the act more as a responsibility than a pleasure. Moral dilemmas in which you feel pulled between sexual attractions and the criticisms of others are possible. If you are a closet homosexual, you may feel pulled between a strong (and obvious to others) attraction to someone you see routinely and the desire to keep your lifestyle and feelings private.

You are more aware of the dark side of human nature. You see evidence of it manifesting both in others and in your own personality. Normally you see individuals as a blend of both good and bad, black and white with variations of gray. But during this year, the separation seems more apparent and the grays tend to fade away. Black and white, good and evil, become all-or-nothing traits, different sides of one personality manifesting themselves at different times. The contrast tends to be stark. Situations you are involved in during the year, conflicts and stress cause the more negative traits to manifest. For example, if you are involved in a love-hate relationship, you will probably see both the best and the worst you and your lover have to offer. The stress of the situation, coupled with conflicting emotions, will tend to foster both strongly positive and strongly negative relating patterns. Disputes over a will or an inheritance can also magnify the split. You may think you know your relatives until either you or they contest a will. In the most extreme cases, you might be harassed by someone who is sadistic or at least has a sadistic side to his or her personality. Personal vendettas are rare but not unheard of.

Use this year to eliminate power structures which are controlling you. Develop psychological self-control. Gain insight into your own personal contributions to your situation. Work towards self-mastery.

SATURN IN THE 9TH HOUSE

Saturn in the 9th house is associated with learning in one form or another. Commonly it can signify a formal education in an established university or college, but learning may instead involve an evening class or correspondence course. As a rule, the course of study is established rather than self-taught. Those who wish to continue their formal education on a high school, college or graduate level tend to experience some minor delays or impediments along the way.

For example, one man who was employed full-time in a newly established office felt he could not start his degree program until he had organized his daily work schedule. After six months, when the office routines became easier to handle, he was able to start school without feeling pressured by job demands.

Perfectionism is associated with the need to learn and individuals look forward to attending school when the time is right. They want to be able to devote their full attention to course work and are willing to wait until outside issues are handled. The delaying characteristic for this placement is especially true if Saturn is placed in the 9th house near the 10th house cusp, and it transits by direct motion out of the 9th during the solar return year. Those who do not wish to enter a degree program may take individual courses to suit their needs, but even these courses will tend to be very structured.

On the other hand, some individuals would rather teach than study with this placement. If this is your goal, you must organize your notes into lesson plans. You can experience some frustration until you have structured your thoughts into a well-defined body of knowledge. Those who have already taught in recent years need to reassess their ability to continue teaching, given certain newly developing situations and possible impediments.

The 9th house is also the house of our beliefs, whether philosophical, religious, or mundane. This is a time when you may realize conflict between what you believe now or have been led to believe in the past, and what you are experiencing in the present. A relationship activity or professional endeavor you are currently involved in may specifically contradict what you once thought to be true. This situation will continually test your beliefs until you come to conclusions which are more consistent with reality. This conflict can occur on higher level of thought as readily as it might relate to more common beliefs. Most frequently, it will encompass and affect both. Inconsistencies which exemplify and fuel your awareness of the conflict constantly arise from everyday situations.

Insight can come from the simplest exchange. The extent to which these insights have ramifications on either higher or mundane thoughts is directly related to the individual's level of consciousness and ability to focus on multidimensional issues. For example, a nurse practitioner dealing with mostly low-income elderly patients in clinic environment was surprised to learn that doctors who prescribed medicine for high blood pressure and heart disease frequently neglected to counsel these individuals about the positive effect dietary control regarding these two illnesses. The nurse practitioner had a strong background and belief in the holistic field and found this lack of attention to the whole individual upsetting, counterproductive and not in keeping with the Hippocratic Oath.

In another example, a lay person who regularly participated as a church assistant doing charity work aimed at helping the poor began to question the pastor's allocation of funds to various projects. Money was being diverted to finance the pastor's need for comfort. This individual spent much of the year pondering his faith, his allegiance to his pastor, and his belief in what the church stood for.

Hypocrisy seems to keynote this Saturn placement. You are acutely aware of what others espouse and whether or not they practice what they preach. You are also acutely aware of those philosophies and beliefs which differ greatly from your own. These philosophical inconsistencies and differences are to help you discriminate what is real and true for your own life path. You must verify your current and developing beliefs within the context of daily experiences and immediate life situations.

Established beliefs form the backbone of the decision-making process. Normally they are meant to have helpful, practical applications to issues, conflicts and situations we face. Those beliefs which dangle in mid-air without any substantiation are useless. Remember, the 9th house is also the house of prejudices, intolerance, and beliefs not based on fact. You could be overly critical of others and their beliefs, setting standards too high for anyone to reach. Saturn indicates that the weeding-out process has begun and a realistic reassessment of all beliefs is called for.

SATURN IN THE 10TH HOUSE

Saturn in the 10th house is equated with career breaks and limitations, some of which are very positive. It is true you can either lose your job or give up your present position during the solar return year, but this turn of events may be desirable. For some individuals, especially for those who are older, this placement signals retirement or leaving the job market entirely. For career mothers, choices focus on quitting work to raise children or continuing work while juggling two demanding schedules. If you have clearly defined personal and professional goals, you need not have any trouble with this placement.

For most people, Saturn in the 10th shows an old career closing down and a job move to something new. Rather than a break in your career, there is a break in your professional momentum. You lose skills and experience while moving from one job, company, or profession to another. In your new position, you must start from scratch. You will be working with new techniques and concepts, and at first, minor duties will take longer and appear more difficult. It takes time to get back into the swing of things, and during your orientation period, you can feel underqualified or inept. Frustration comes from not knowing what to do and the tendency to make mistakes, which is common for newcomers.

If you quit your present job to become self-employed at this time, you may go down several dead ends before you develop, establish and maintain a routine and clientele. If you are creating a job from nothing, you are starting from scratch. The same is true if you are self-employed and must relocate. Success in a new position takes time and effort, but it can be done by those who work hard. The rewards will be clearer at the end of the year as the old chart fades away.

If you are comfortable with your present position and have always been a hard worker, you may be asked to take on more responsibility at this time. The natural tendency with this placement is to assume personal responsibility in the career area. A promotion to a managerial position or an increased workload are two ways to do this.

Serious problems associated with this position occur when you refuse to assume any responsibility in your career or for problems on the job. Suppose you do not like your job but refuse to contemplate any changes. Limiting your career options in this way creates a strong feeling of frustration, restriction and monotony. If your present job is unfulfilling and unrewarding, you will not be able to work to your potential. Perhaps you should quit rather than stagnate. Those who refuse to reassess career options and take responsibility for their own career decisions, or lack thereof, end up going nowhere while Saturn is in the house of destiny, career and long-term patterns of growth. Individuals who make changes and start from scratch learn to work with job frustrations to structure a new professional environment. This is the goal whether you stay in your present position or move on. You must reassess where you are and where you are headed professionally. Map out the practical steps necessary for the journey.

SATURN IN THE 11TH HOUSE

Saturn in the 11th house indicates a need to reassess goals for the future. The goals you presently hold are no longer practical in light of new situations you are moving into or new information you are receiving. They are outdated either because of external changes or your environment or internal changes in your personality. Perhaps they are now inconsistent with your present or future needs for fulfillment. A lack or loss of goals can be implied by Saturn in the 11th house, but generally as old goals prove unworkable, new ones will arise to take their place. It is only rarely that the individual decides to take a year's vacation. Working mothers who take a year's leave of absence from work to care for a child sometimes have this placement.

Now is the time to rethink your future and take corrective action where necessary. A new college graduate, also a new army wife, followed her husband to Germany where she discovered she was not allowed to work in her chosen profession despite her degree. She had to rethink her goals for the coming year while overseas. Those who are newly divorced must also rethink their goals in light of their single lifestyle. This can be a year of great accomplishment for those who settle on a direction quickly and move steadily towards achievement. The new goals you develop tend to suit you better than the old ones, but they need a great deal of work to become a reality. Saturn is the ruler of hard work, practical applications and realistic ventures.

Working with a group of people may give you a better sense of discipline and organization than you would have on your own since you are more likely to take on extra responsibility. You can work harder for the sake of group goals and the extra push might prove ultimately beneficial for all. One astrologer with this Saturn placement taught a group of other astrologers a particular astrological technique and eventually used his organized notes to publish a complete course. However, group dynamics can be cumbersome. You might feel that your individual opinions and identity are lost because of the democratic process and the need to conform. There will be times when you disagree with group leadership or direction. Even if you are the leader, your ideas and opinions can still be watered down by the need for agreement. But working with a group might help you to get off the ground and push you toward making your goals a reality.

Friendships can be lost during the year and there are several possible reasons why this might occur. Most commonly, either you or your friend move, relocating at a great distance away. If you and your husband, wife, lover or partner are splitting up, friends will tend to fall into either camp. They drop away if they were primarily connected to your mate or set on seeing the two of you as a couple. Serious or older friends are also indicated by this placement. Those friends who were mere playmates tend to be less prominent as the year wears on. Your focus is more on business relationships geared towards making connections than on casual friendships. Obligatory friendships (more commonly seen in the business world) are possible.

Since the 11th house is the 2nd from the 10th, money from your career may decrease this year, especially if you are pursuing a goal that cannot be financially lucrative, at least in less than one year's time. Long-term goals tend to be more consistent with this placement than short-term goals or immediate compensation. In this sense, you may work harder for the same amount of money or even less. Wage freezes are rare but possible; raises may be delayed during the year. For any promotion, you may have to assume extra responsibility to receive any monetary increase. Volunteer work can also be implied by this placement.

SATURN IN THE 12TH HOUSE

Saturn in the 12th house implies the presence of unconscious blockages which must be overcome. Failure to overcome these inhibitions will lead to limitations in what would usually be considered normal activity. The emotions governing these limitations are fear and guilt, and breaking free generally involves much anxiety. You must face your own fears before you can go on with the maturation process, since presently they prevent you from branching out and taking risks when you should develop greater freedom of movement. Frequent fear issues include but are not limited to: fears of not being loved or being rejected; fears of inadequacy with a constant need for reassurance; fears concerning sexual performance or orientation; and fears of losing control or being overpowered. Many times the fear-producing issue is ill-defined and ill-founded but supported by self-defeating attitudes, free-floating anxiety and heightened vulnerability. For these reasons, fears associated with psychic impressions are not unheard of.

Guilt is the other most common unconscious restriction. There may be certain things you will do, or other things you will not do, simply because of the guilt you will feel one way or the other. If you are being controlled by guilt and do not investigate or understand these feelings, you will continue to feel forced to behave in certain ways. For example, a middle-aged woman who lived with and cared for her elderly mother found she was both tied down and drained by her mother's constant need for care and attention. But she repeatedly refused to seek out a part-time nurse or companion to stay with her mother while she pursued her own interests. Every time she left her mother with other family members, she would return to find the older woman somewhat disoriented and confused. She would interpret this disorientation as her own fault and feel guilty for leaving. Eventually, she became totally tied down by her mother's unconscious signals for attention.

Saturn in the 12th is the astrological equivalent of a responsibility vacuum cleaner. There is the tendency to feel responsible (and subsequently guilty) for everything that goes wrong. You will tend not to delegate authority easily and will try to do everything yourself, perfectly, all the time. In short, you overcompensate for real or imagined shortcomings. Only by facing your fears and confronting your feelings of guilt can you begin to live a more normal lifestyle.

There is a lack of "reality presence" with the placement of Saturn in this house. Saturn rules reality, but here in the 12th house (normally ruled by Neptune) reality seems to lose some of its clarity. Truth is more elusive. The fears, guilt and unconscious inhibitions normally associated with this placement are partly caused by an inability to discern what is real and what is fantasy. Here there is no objective or even subjective reality, only vague impressions of what is presumed to be real. This makes the going tough. Without a clear base to work from, it becomes easier to worry about future possibilities and past mistakes. Consequently, fears and guilt tend to prosper and grow at this time.

The main goal of this house placement is to come in contact with the weaker, more irrational parts of your personality which inhibit your progress. This is easier to do if you are in the process of being thwarted or feeling afraid. Only by experiencing the frustration and fear inherent in the conflict, will you actively seek ways to overcome the blockages and resolve the associated issues or problems, so life can begin to return to normal. It is to your advantage to move towards the fear rather than retreat. Investigate the issues which seem to block your progress. Look for alternatives. The tendency is to miss the obvious or not understand what is normal under the given conditions. Only by discussing your situation with others and looking for new options can you alleviate the pressure. Unfortunately, you will be doing this while feeling your most vulnerable. But you must analyze your behavior and question those actions that you do not understand. Fear and guilt are sure signals that you need to investigate unconscious blockages and make adjustments in the way you are living and handling issues.

Saturn in the 12th can also show strong obligations that cannot be avoided, or the need to become responsible for someone less fortunate. Couples with newborn babies and children with elderly parents to care for frequently have this placement. Family members in need must be cared for, but friends in trouble or those less fortunate can also be the object of your concern. The need to behave in a responsible manner includes the desire to care for someone who is unable to be responsible for him or herself or needs to be saved from a difficult situation. Occasionally, the person you are most concerned about is hospitalized during the year. The need to care and the need to face a fear are usually intertwined in some way. Sometimes, the person you care most for is the person you fear. If so, by giving, you receive; by protecting, you become less vulnerable; by reaching out, you cross the bridge from fear to understanding.

If Saturn is in the 12th house but close to the Ascendant, tasks and responsibilities will tend to become stronger and more visible as Saturn crosses the Ascendant by transit and moves into the 1st house.

CHAPTER NINE

URANUS IN THE SOLAR RETURN CHART

Introduction

The rotation of Uranus around the solar return chart mimics that of the Sun. Both generally move 3 houses clockwise each succeeding year for those who remain in the same location; however, because Uranus is a transiting planet while technically the Sun never changes degrees, Uranus slips more quickly from angular house progression, into succedent houses, followed by cadent. Therefore, the total passage of Uranus through all the houses takes approximately fourteen years, a much shorter cycle than the longer, more repetitive cycle of the Sun (thirty-three years). As the Sun passes through the angular, succedent, and then cadent houses, the tendency is to repeat the specific house placements two or three times at four year intervals. On the other hand, Uranus placements are usually not repeated.

For example, the Sun in the 1st house this year will most likely move to the 10th next year, then the 7th, 4th, 1st, 10th, 7th, 4th, 1st, 10th, and 7th before passing into a succedent house. It ultimately takes the Sun about ten to twelve years (depending on signs of long or short ascension), before it passes on to the next house type (angular, succedent, or cadent). This length of time allows for repeating Sun placements.

Within a similar span of time (approximately fourteen years), Uranus would have made a complete cycle through all of the houses. Suppose Uranus were in the 1st house this year; chances are it would fall in the 10th house next year, then the 7th, 4th, 2nd, 11th, 8th, 5th, 3rd, 12th, 9th, 6th, and 4th. These positions are approximate, and again, placements are subject to variations caused by signs of long and short ascension and the transiting speed of Uranus.

The distinction is obvious. The Sun represents a conscious movement toward change through growth and choice. The individual's focus is repeated after four years, and transformations may take a series of repetitions before maximum growth can be realized. On the other hand, with Uranus the tendency is not to repeat since usually there is no second chance four years later. Most likely Uranus will slip further counterclockwise into a new house mode on the next swing, so it is important to move with the winds of change when first they come calling and make the necessary transition at that time. By studying the placement of Uranus, you can become aware of when changes are more likely to be made quickly (one year placement in a house only), or when they are more likely to be made in a two-stage process (when Uranus returns to the same house four years later). You will also be able to understand the pattern of growth through the cycle of change.

Uranus is commonly seen as an indicator of change, and this is true for the solar return placements also. Conditions associated with the house position of Uranus are likely to evolve significantly over the year. Major changes or incessant fluctuations emerge as the pattern of manifestation, depending on how the individual copes with his or her situation and needs. Transformations may occur quickly and require a long period of adjustment, or progress slowly, possibly occurring late in the year and only after a long period of anticipation or restlessness. As a rule, most changes are expected, predicted, and initiated by the native him or herself. Many are carefully planned and well executed, involving a minimum of tension and anxiety. When working positively with the Uranian principle, the key is in the restlessness. Those who are very attuned to their own need for growth will feel the restlessness arising in outgrown situations and respond accordingly. They will welcome the opportunity for adjustment, and make all the necessary preparations for the coming change. They will not lock themselves into one particular situation or pattern of growth, but will allow changes and insights to develop and evolve naturally along the way.

Generally, it is only when the individual thwarts his or her own desires or tightly controls situations that tensions manifest in the form of anxiety and nervousness. This anxiety results from ambivalent feelings which develop in those who resist needed changes. These people get caught between their own fear of change and a strong desire for the very change being thwarted. An ambivalent mind-set takes over, one built on an approach to, and also an avoidance of, a set goal or change. Remember, Uranus is associated with conscious and unconscious choices for change. Its appearance somewhere in the native's solar return chart implies the area of life in which the individual desires to progress through change. When one is unable to face or effect needed and desired changes, the mind splits between two mutually exclusive goals and anxiety results. Eventually, the mental ambivalence and erratic commitments to two very different paths are reflected in fluctuating external conditions. Long overdue, yet still avoided changes, tend to manifest in the environment as disruption caused by others, or by neglected areas of attention.

Adding to the anxiety is a perceived loss of control over external situations. Eventually, the restlessness and tension rise to a feverish pitch until one finally agrees to make changes, or can no longer prevent their occurrence. When conditions reach this intensity, individuals usually make reactive changes, without careful consideration or adequate preparation. Sudden upheavals occur rather than welcomed transitions. If we would truly be in control of our own destiny, we should listen to the need for change and respond to the earliest hints of restlessness. We should give ourselves the freedom to work towards a conscious transition before a crisis arises.

Consistent with the desire for change is the need for freedom. You cannot maneuver if you are locked into a restrictive environment. Sometimes the push for freedom is a prerequisite for change. At other times, the change itself becomes the motivating force behind the process. Occasionally, both mechanisms are operating. For example, you may have to convince your boss to give you the freedom to make needed changes in the daily office routine; at the same time, the changes you make could streamline procedures, creating more leeway in your work schedule.

Freedom allows the process of change to occur smoothly. When we are functioning at our best and working positively with the

Uranian concept, we move easily through a series of attractions, detachments and independent actions. We learn and grow from each of these encounters. While in the process of attraction, we are drawn to certain situations, persons or concepts. It is important to observe what experiences we are drawn to because they indicate conscious or unconscious needs and patterns of growth. By accepting different situations, interacting with different individuals, or listening to new ideas, we take in new information and learn to value various facets of life. We see that no one person, place or idea is perfect. The same is true of our existing circumstances, which we hold onto so tightly.

With this insight begins the period of detachment from those situations, persons, or concepts which restrict our growth or no longer have anything to teach us. We are able to step back and intellectually reassess our involvement and commitment. We can either choose to separate physically from difficult situations, or take the corrective action needed to adjust present circumstances to our needs.

Uranus does not always imply a complete and total separation. We are capable of detaching from the original attraction while still maintaining the experience, relationship, or train of thought. It is the independent action we ultimately take which defines the degree of separation.

We learn as much from separation as we do from attraction when we define what we don't like as well as what we do like. The changing environment, the comparisons and contrasts, accentuate what is important and what is not. One should remember that this process of change through attachment, detachment and independent action is ongoing, occurring on many different levels simultaneously. The process can be either quick, occurring numerous times in one day, or drawn out, requiring a year's time span. For example, suppose you are building a new home. This is a year-long project and during this time you will be drawn to many different construction plans, ideas, and subcontractors. But as you begin to work with the possibilities you will accept, reject, or change options to suit your needs. This process takes place over months, but also simultaneously many times in one day.

The exposure to various ideas, situations and people stimulates creative thought through sudden insight into changing situations. All planets represent a creative process. Venus is the planet of creativity experienced through beauty, while Uranus is the process experienced through change. Because of the changing panorama of people, ideas, and situations, one becomes accustomed to looking at life from different perspectives. The multifaceted approach encourages the mind to create still newer ideas. In this way the individual begins to participate in the process of attachment, detachment and independent action by creating his or her own original options, and change becomes a process of perception, integrated into the pattern of growth.

Planets Aspecting Uranus

For Uranus' aspects to the personal planets, refer to the earlier chapters in this book. For example, Uranus in aspect to Venus would be given in Chapter 5, Venus in the Solar Return Chart.

Uranus in the Solar Return Houses

URANUS IN THE 1ST HOUSE

Uranus in the 1st house of the solar return is usually indicative of strong and dramatic transitions. It is very likely that you will consider making a major change in your lifestyle, location or even appearance during the coming year. The transformations that arise from the restlessness associated with the 1st house Uranus can be far-reaching and affect many other areas of life. If you are aware of your own restlessness, you can not only welcome the coming changes, but initiate them yourself. This is meant to be an exciting time, though somewhat nerve-wracking. Generally major changes are the norm, but occasionally minor changes suffice. Major changes include but, are not limited to, relocations across country or at least across state lines, lifestyle changes (divorce, living alone or out on your own, communal arrangements or living together, stopping work, starting work, or returning to school), and individual changes associated with personality characteristics or appearance (increased assertiveness, greater demands for independence, and a noticeable weight loss or gain).

You can expect a change of pace if your current lifestyle no longer meets your needs and you feel bored, restricted or restless. It's time for a fresh start and it is to your advantage to take risks and make the necessary changes. Assess the solar return position of Saturn in relationship to Uranus for an understanding of the kind of risk you are willing to take, and what realistic criteria must be met for you to make changes easily. If Saturn is prominent in the solar return chart, you are more apt to take calculated risks, planning your moves carefully rather than jumping into situations quickly. This should be reassuring to those individuals who find it more threatening to deal with Uranian impulses.

Most changes center around your need and ability to act independently. There is a strong desire for freedom of action and you perform not as a corporate player, but as a free agent. The activities you are involved in may require you to function separately, without the assistance or agreement of significant others. Some individuals seek or finalize a divorce during the year and sever existing ties completely. Others form new, exciting or unconventional relationships. Although you may be relationship-oriented, you may not necessarily be partnership-oriented. If other factors in the chart so indicate, an extramarital affair is possible, but this is certainly not the norm. The predominant need is to take action on your own and for your own sake. In at least one area of your life, you must make your own decisions and function as a separate unit. Nearly everyone can do this while still maintaining existing partnerships, but it is important that those you are involved with give you enough freedom to grow and change. Now is the time to get in touch with your own unique individuality, and to do this, you may have to act differently from others or differently from the way you have acted in the past, breaking old patterns of behavior. Realize that you will probably not adhere to your old standard of behavior. Those who are particularly Uranian may not adhere to social norms either. Personality experimentation is possible and can help get you moving.

Your behavior may get erratic or downright disruptive. If others around you are not adequately prepared for the changes you are making, they will see you as undependable, unpredictable or even out of control. Understand that others may see this whole process as lacking stability. Your concentration can be easily interrupted, and you may find it difficult to work consistently on a project, preferring to work during strong bursts of energy. Uranus does not represent a steady pattern of energy use, but an erratic pulse that moves in fits and starts towards growth and change. Move when the restless energy is there and rest when it is not. Work with the internal process and do not get trapped into a rigid schedule. Activities will not be well-planned since spontaneity seems to be the norm and plans you do formulate tend to get changed.

If you are really rut-bound and fear making any changes at all, the people around you will make the changes for you. Your life can become disrupted by others, especially if you are trying to remain the same while buffeted by the winds of change. Surprises and unexpected upheavals will be the norm as life gets unpredictable. This is a time for a flexible schedule. Own your own restlessness and welcome positive changes.

URANUS IN THE 2ND HOUSE

Uranus in the 2nd shows a change in your earning power and financial situation. Alterations can be major or minor, fluctuating or one-time single events. The adjustment can be either positive or negative, but both manifestations can occur at different times during the year. Positive interpretations include pay raises or large windfalls. Those who have not been working can find employment and those who have been doing volunteer work can now get paid for the work they do. Being self-employed or working on a commission or incentive basis is common with this placement. You might land that big account, exceed previous sales records, or successfully start or expand your own business during the year.

Negative interpretations include a total break in income because of either a job change, relocation, or leave of absence without pay (the most likely reason for this being maternity leave). Those who work steadily may receive or accept a pay cut. Some actually request a cutback in hours (and therefore pay) because of shifting goals or pressing personal needs. It is possible to give up a job with a very good income. Salespersons working on a commission basis may find it more difficult to sell products for one reason or another. Changes in earning power seem to be more closely associated with this placement than disruptive spending practices; however, impulse buying may be a problem for those with this tendency.

Economic change is easy to foresee with this Uranus placement, but it is difficult to note whether the changes will be positive or negative. The aspects to Uranus are not a true measure of the shift. One can earn more with difficult aspects and one can earn less with softer aspects. What is shown by the aspects is the individual's innate desire to foster a financial change and the other areas of life that will be affected. The house placement and the aspects to Venus may give you a clue. Also check progressions and transits. Solar return "work house" placements (10th and 6th houses) can be important since they are a measure of job satisfaction. Any planets in the 2nd house in addition to Uranus should also be considered. The best course of action is to prepare for a rainy day. It is better to practice financial restraint than assume more money than you have.

Uranus in the 2nd house also indicates that values and morals are changing. Nonmaterialistic attitudes are the norm unless there are contraindications in the solar return chart (e.g., earth emphasis, prominent Venus, etc.). As a rule, inner fulfillment and personal satisfaction are more important than material possessions. Money itself does not matter unless the financial situation is dire. Possessions will not be important and this is a good time to unload useless things. During the year, you may not be interested in purchasing new items or replacing old ones. If you are not aware of your inner needs for emotional rather than material satisfaction, you may only experience the financially difficult side of this Uranus placement, especially if you repeatedly attempt to gain material wealth and acquire new things despite emotional considerations to the contrary.

Morals are also in a state of flux at this time. It is best to allow for review and change. Regardless of changes, you may deviate from your own code of behavior. Sometime during the year, you may be asked to defend your position if your actions or values conflict with the standards of others. This will give you an opportunity to better define your own priorities and standards.

URANUS IN THE 3RD HOUSE

Uranus in the 3rd house is associated with rapid thinking processes and changing thought patterns. The mental acceleration can work to your advantage if you are in a pressured situation where decisions must be made quickly or information must be learned in a short period of time. This is an excellent time to take a crash course or intensive which forces you to learn a lot of material in a fast-paced classroom situation. Independent learning and correspondence courses are also possible since you have the ability to study on your own and move ahead at your own pace. The information under study should be unusual in some way, geared towards stimulating creativity, insight and free thinking. Do not limit your thoughts because of preconceived notions.

It is very likely that during this time you will receive or learn information which could have a very strong effect on your perception of reality and your pattern of thought. This information may come to you through your studies, through conflict with others, or through intuitive insights, but the effect of the information will be transformational, causing your perception of reality to go through a period of adjustment.

During the year, your opinions will change sequentially as the weeks go by; therefore, this is not the time to make irrevocable statements. You may end up eating your own words later on. A better course of action is to allow the mind complete intellectual freedom for the purposes of investigation and insight. Do not limit your thoughts to what is already known, what is rational, or what you are told. Allow your thoughts to form intuitively, creatively or unconsciously. Work with information from many levels of experience to gain a better understanding of your true reality. You can gain great insight if you read between the lines and accept all information that comes your way without judgment. Let the final say occur toward the end of the year after all the information is in.

There is a propensity for anxiety and nervousness with Uranus in the 3rd house. Mental problems are associated with a fixation on what is desired or expected rather than an openness to new possibilities for the future. Anyone who experiences acute anxiety or panic attacks should seek counseling. But most individuals will find this placement somewhat nerve-wracking unless they are able to channel mental tension into productive pursuits such as learning, insight, creativity, etc. A lot of the tension arises from the sequential changes in thinking that occur during the year, but daily disruptions, constant interruptions and hectic schedules also contribute. It is difficult to plan a routine while Uranus is in the 3rd house. At the very least, you may be attempting to do too much in too little time. The tendency is to work with scattered thoughts and many activities, scattering your energy in several directions even on a single day.

Crisis situations rather than well-organized routines are the norm, since your daily activities generally demand more time than you actually have available. If you are trying to do five different things on any one day, and realistically you can only schedule in four, each day will have a different rotating schedule so that you can fit everything in on a weekly basis. Your hectic schedule leaves little room for adjustments when problems occur, and of course, problems do occur. Imagine ad-libbing your daily routine for a year because you and everyone and everything around you are changing. Daily life just does not run smoothly enough for you to stick to a schedule.

Uranus can be indicative of original thinking and creativity. You need time to create. Ideas can be close to the surface, yet just out of reach. This is why disruptions can be especially annoying when they interrupt the creative process. It is important to establish some quiet time to allow insights to manifest. Find a way to retain your thoughts and insights. If you are on the go a lot and creative insights are developing quite rapidly, you can use a tape recorder to preserve your thoughts. Allow erratic energy to flow out freely and come to the surface. This is not the time to limit your mental wanderings simply because you question their logical implications. The thought processes are not very organized or directed this year. There is the tendency to go around and around to get to a simple truth. Play with ideas and concepts. Restricting your erratic insights could mean limiting your creativity.

Conflicts are not directly associated with Uranus in the 3rd house, but a bad temper can be. The stress of daily life can make you snap at people and make rash statements you later regret. The connotation here is more akin to, "Open mouth, insert foot," than argumentativeness, but you can only make *faux pas* so many times before you enter a verbal battlefield. Learn to think before your open your mouth. Do not jump to conclusions. Remember that anything you say can get back to others and can be used against you.

Problems with brothers, sisters or neighbors may arise during the year. These people may not be especially dependable or predictable and a conflict of interest may occur. This is not the more common interpretation for this placement. The emphasis seems to be more on the mental changes and creativity. You can work things out through discussions but be careful what you say. Neighborhood problems can be addressed through involvement in the community association. It is possible that rather than disagreeing with a neighbor, a close friend in the neighborhood needs your help during the coming year.

URANUS IN THE 4TH HOUSE

Uranus in the 4th house indicates that your domestic life is very unsettled and some disruptive change occurs in the home. There are several reasons possible for this disruption. Individuals often move (sometimes repeatedly) or try to move during the year. The actual move, if it occurs, may involve a major relocation. Preparations for the sale of the old house and renovations in the new house can drag on for months. Those who do not choose to move during the year may decide to rearrange the house or remodel part or all of the present residence. Some build on an addition. In general, major renovations (the kind where people switch the living room and the kitchen) are more likely than mild redecoration. Sudden repairs are possible and often freaky things can happen. In one case a chimney fell down, and in another situation the house literally moved on its foundation.

The actual house may remain the same but the number of occupants living with you could change as others come and go during the year. A child may return from college, or choose to live elsewhere. Adult sons and daughters, elderly parents, or roommates may move in or out, either temporarily or permanently. The coming and going seems to interrupt the tranquility and routine. At the very least, if nothing else changes, you will tend to be restless when at home and may not spend much time there. You may travel, or live with others and be in and out sporadically. Home may not really feel like home and you may feel uprooted much of the time.

Besides domestic disruptions, emotional disruptions are also possible. In fact, the greater the domestic change, the greater the transformation that will be occurring on an emotional level. If you do not understand the transition, you may be moody or detached. You may not trust others with your feelings, especially if they have a history of being emotionally unpredictable and undependable, leaving you to feel that your expectations will not be met if you approach these people for either support or comfort. If you cannot get what you need from others, vocalize your dissatisfaction, but concentrate on verbal communication to get your point across.

Realize that others may have grievances against you also. Friends, family and lovers may see you as emotionally unpredictable and undependable. You may not be aware that you are behaving in a way that conflicts with your own need for security and safety. Your ability to make a commitment will change back and forth and you cannot establish what kind of emotional security you want to have in a relationship until you decide what you are capable of contributing yourself. If you are unable to reach an understanding, seek support elsewhere or your discontent will settle into grouchiness, anger, and manipulation. In very negative situations, you can distance yourself from family members or those you live with. Fights, tensions and disagreements become more likely as negotiations break down, and can lead to separations and broken ties.

Less common possibilities with Uranus in the 4th house include changes in the health and/or independence of family members. Unexpected illnesses are rare, but do occur, especially in elderly parents or grandparents. When they do occur, illnesses tend to come on suddenly and exhibit an acute stage which is usually temporary. It is during this time that the sick person will need assistance and may actually move in with you. Moreover, parents or children may not be able to function independently at this time for reasons other than illness. Surprise pregnancies, motherhood, unemployment, job relocation, etc., also change one's ability to be independent. Learning to develop a sense of freedom in the home environment is associated with this placement.

URANUS IN THE 5TH HOUSE

While Uranus is in the 5th house, you want to be able to express yourself freely. You may need to function independently of peer pressure and relationship demands in order to do this. Your personality style is changing and you may take on more Aquarian characteristics. It is only through the change, and the freedom and independence that foster it, that the uniqueness of the individual can emerge. You must limit the influence of others to search for the identity within. It is important that you use this time to be your own person, one of a kind. You do not want to pattern yourself or your behavior after someone else. You can have something different to contribute to the environment.

This is not to say that others will find the transition from the "old you" to the "new you" easy. Depending on the restrictions and expectations others place on you (and which you allow to exist), this can be a difficult year or a very easy one. The changes in self-expression might cause conflicts with significant others if they do not believe in what you are trying to accomplish or who you are trying to become. These people will need reassurance.

If you feel very limited and restricted by others, you might think you have to be very rebellious and contrary to break their hold. Generally, assertive independence is all that is required. You need not contribute to the conflict. This is the year you will want to change all personality habits that are inhibiting self-expression. Be mindful of the ways you compare or contrast to others and pay special attention to the ways in which you are different. Differences matter this year and they mark growth. Allow yourself the freedom to flow with those distinctions which make you a unique human being.

This can be a very creative year, especially if you are already involved in an artistic field, but creativity need not be limited to artistic endeavors. Uranus represents the genius, the innovator, the inventor and the individualist. This is a time when strong individualism enhances the ability for original thought. You grow to see things differently as the year progresses and it is most likely that you will have to deal with a creative problem or issue in a new way. The more you allow your mind to float free, the greater your ability to think up new ideas. This is a great placement for the free thinker, writer, student researcher, artist or craftsperson.

The only difficulty associated with creativity and this Uranus placement is the tendency to go through a short frustrating period of transition. Most likely, there will come a time during the year when blockages occur because the creative style is in such a state of flux. These blockages are not permanent, but serve as a signal that creative shifts are now taking place and there can be a transition to a much higher level of attainment in artistry for those who understand the transformation and go with the new energy fearlessly. Do not be dismayed by this development. Go with the flow and trust that new skills await those who can progress. A prolonged blockage shows a resistance to new forms of expression waiting to be born. Adjustments may take a while, but they are worth the effort.

Sudden attractions are possible during the year, but not the norm. On-again, off-again episodes, either in a new unbonded relationship or an existing one, are common. The person you are involved with might not be dependable, could live in another area, or might be unable to make a greater commitment at this time. Relationship breaks, for one reason or another, are likely, and in general, the relationship will not run smoothly. You will not settle into the "boy meets high school sweetheart and dates high school sweetheart exclusively" routine. Who you are attracted to may surprise you. Potential lovers may be very different from those you have responded to in the past. They could be free-spirited individuals, having few restraints. These people are probably representative of your own need for freedom of self-expression.

If you are attracted to someone who is very conservative and limiting, perhaps you find it necessary to rage against the limits. Fear-

of-freedom issues are likely. Either way, whether you pick someo
far out or very straight, you tend to draw those who are an extren
of some personality trait you are trying to cope with. Unbonde
relationships this year are not only erratic, they also tend to b
mismatched pairings. Even when two very similar people get toget
er, the emphasis will be on forming a unique relationship which i
some way breaks relating patterns of the past.

If you have children living at home, they will be more indepe
dent, unpredictable, disruptive or unusual during the year. This ca
be a time when they are experiencing great changes in their lives an
these changes can be unsettling. The most common change involv
relocation. Either your family moves, or others move away, leavin
your child without a best friend. New schools, puberty, and additio
to the family are some of the other changes affecting children a
their behavior. Any change can put pressure on kids to adjust, an
Uranus in the 5th house usually signals a period of adjustment. A
your children make their way through the transition, you will be le
able to predict their responses. For this reason, it may be wise to kee
a close watch over their activities, especially if they are young. The
may be ready for greater independence, but still in need of yo
assistance and advice.

Disruptions can come in the form of behavior problems or mino
illnesses. If they are stressed, children are less likely to "perform" a
more likely to act out or become sick (colds, ear infections, and flu a
common for young children). Expect schedule changes. Realize tha
disruptions of any kind show a need for more attention from yo
Although you may be less patient at these times, understanding a
a calm attitude will work more to your advantage.

Unusual learning characteristics in children can require spec
attention at school. Your children might need individualized educ
tional assistance in one or more areas, such as a remedial or gifte
course. Those parents with older children will notice a strong push f
independence. Those who are still living at home may decide to mo
out and those who are already on their own may move away. Only i
very negative situations will grown children be very disruptive
erratic. The real need is for the child to establish an independent a
unique identity.

URANUS IN THE 6TH HOUSE

Uranus in the 6th house can show a job change during the year or a change in working conditions. Changes could be either self-initiated or beyond your control. Temporary working positions or breaks in employment are possible. For example, one woman worked for a nonprofit organization. Her salary was paid for by a government grant and when the grant money ran out, she was let go, only to return when new funds came in. If you are presently working for someone other than yourself and dissatisfied with your job, now is the time to investigate new placements. You can change companies entirely, but it is possible simply to transfer to another office under your present employer. Some individuals will ask to work on independent projects of their own choosing, or go one step further and create their own businesses, becoming self-employed.

If you maintain your old job, changes in your daily routine are likely and can involve relocation of the office, temporary change of duty station (travel), new office procedures, or the installation of new computer equipment. During the period of transition, simple procedures become very complicated and normal operations will be disrupted. In very negative situations, especially those involving conflict and even sabotage, work can come to a standstill.

Scheduling changes are also possible during the year. Your office adopts a flexible or rotating schedule, and you might be able to set your own hours. If there is no change in your job at all, work can become very nerve-wracking, tedious and stressful, particularly if you are restless and need a change of pace. You will be easily bored with repetitious tasks and needless restrictions. If all goes well, any change in your job will involve a variety of tasks and greater freedom. You need a position that gives you a new challenge, coupled with a changing schedule and the freedom to work to your potential.

Trying to stay at the same job doing the same thing in the same way will create stress. Changes are not only likely, but necessary for progress to occur. Until adjustments are accomplished and your office streamlined and better organized, your normal workday can be constantly disrupted by little crises. This is also true if you own your own business. You should be contemplating adjustments which will eventually lead to a more efficiently run business. The disruptions may occur during the period of transition to the new procedures or may actually prompt the procedure changes.

All this stress and tension at work can affect your health if you do not protect yourself. Your physical and emotional well-being is directly tied into your job situation and should be watched carefully, especially during a time of stress. Learn relaxation techniques and use them to rid your body of tension at the end of the day. Do not bring work or problems home with you. You need your free time for recuperation. Cut back on caffeine drinks and take stress-preventative vitamins when necessary. Most importantly, correct those job situations which are the most stressful.

Health habits tend to be erratic during the year, and consequently you may experience sporadic health problems. A lot depends on age. Generally the younger and healthier you are, the less likely you are to be sick. Tensions at work are the most likely cause of stress-related illnesses. Work demands can disrupt health routines, especially if you must travel for your job or work crazy hours. Wide variations in your eating habits (including rapid weight losses or gains) can occur. Stressful work habits or rotating sleeping patterns are not in your best interest. Practice moderation in all things. This is a good time to change your health regimen consciously in order to foster a healthier body.

URANUS IN THE 7TH HOUSE

The primary interpretation for Uranus in the 7th house focuses on changes to relationships. Depending upon what you are used to, relationships can change in a number of different ways. If you are not in a major relationship or have not been up until this point, this can be a milestone year for you, one in which you change your style of relating and push for greater intimacy. Sudden attractions are possible and the person you have your eye on may be quite different from what you would normally expect. Attractions can be very exciting, but also unpredictable. Patterns of relating tend not to be soothing, but somewhat disruptive. It may be difficult for you to depend on the person you are involved with, since he or she will not want to be tied down at this time. Freedom is an issue for both of you and togetherness may be on-again, off-again. Marriage is probably not an option during this year, though you may live together.

Existing relationships may go through a period of transition because one of you wishes to make a major change or needs more freedom of movement. Usually the freedom to make a major change is sufficient, but sometimes the push is stronger. Although separation may be indicated by Uranus in the 7th house, it is more commonly associated with Uranus in the 4th house square to the Ascendant-Descendant axis. Separations in consciousness are frequent with Uranus in the 7th house. If your partner is working on a major project, he or she may seem detached and preoccupied, or extensive travel may be a job requirement, taking him or her away for part of the year.

On the other hand, you might be the preoccupied or busy person. It is common to be separated for short periods of time for one reason or another. Perhaps one of you is married, lives out of town, or is busy with school, work or other activities. Changes directly affecting you can come through the spouse, lover, or business partner. For example, the person you are living with relocates and you must decide to move also or separate. It is also possible for you to instigate changes on your own.

In both new and old relationships, a lack of true intimacy is a frequent complaint. Connections might seem distant, commitments erratic. A conflict of interest is possible, especially if there are squares or oppositions to Uranus in the 7th house. It is normal to experience some relationship oscillation during the year. Changes sometimes necessitate a disruption of the intimacy routine and flow, but some couples move closer as they grant each other greater freedom of movement or expression. This is a time to become a friend to the person you are closest to. Distances will be more evident and serious in difficult relationships, but good relationships will survive and grow from the transition.

URANUS IN THE 8TH HOUSE

Changing financial conditions are common with Uranus in the 8th house. Usually the change directly affects shared resources, but sometimes the change also or instead affects debts either positively or negatively. Desired or not, the hallmark of this placement is financial independence either from the partner, some other person, or an overwhelming debt which is restrictive. It is possible for shared resources either to increase or decrease, but generally there is a break in the partner's income during the year. He or she may quit work entirely or switch from one job to the next with a period of unemployment in between. Problems may be very minor and only situational. For instance, if your spouse is in the service working overseas and you are employed in the States, you may choose to work and live entirely off your own salary, not dependent on other funds at all. This will allow your spouse to save his or her salary for future expenses. The money is not lost permanently, only withdrawn for one year's time. Your financial situation may change specifically because of others you depend on, especially if there are oppositions from the 2nd house to the 8th house Uranus.

Unfortunately, some couples choose to fight over money and possessions during this year. Problems arise when one of you refuses to share expenses or funds. The positive manifestation is to move towards financial independence by separating accounts and responsibilities, but some people turn this process into a battle scene. Serious situations involving separation and/or divorce can eventually end up in court. Decisions, though fair, may not be worth the trauma you go through. If you see serious problems arising, use this time to separate your funds and protect your resources by dividing accounts, funds and expenses.

Debts either grow dramatically or are completely wiped out, but they do not stay the same. Increases result more from necessity than luxury. The most common manifestation is to eliminate all debts entirely either by paying them all off or by declaring bankruptcy. The need for financial independence extends to the debt situation and you do not want to be bothered by large bills or overwhelming and restrictive debts. Sometimes your future plans require a debt-free lifestyle.

Changing attitudes may affect your sex life. The issues of sexual freedom or sexual preference (homosexuality, bisexuality or heterosexuality) may be important either to you or someone you are close to. During the year, you may experiment with new techniques or change old practices which you no longer enjoy or benefit from. In the age of AIDS and herpes, many with this placement have begun practicing safe sex or curbing their appetites. A very negative manifestation might be a disrupted sex life, one in which encounters are on-again, off-again, brief exchanges or twisted experiences that are not meant to be fulfilling.

Deaths are not likely to occur during the year, but if one does, it will be unexpected and the manner of death will involve unusual circumstances. This is a very rare possibility.

Your insight on the psychic and/or psychological level may increase dramatically. You are better able to perceive the motivation and manipulations of others, especially in financial and sexual situations. Often these perceptions will come in the form of sudden

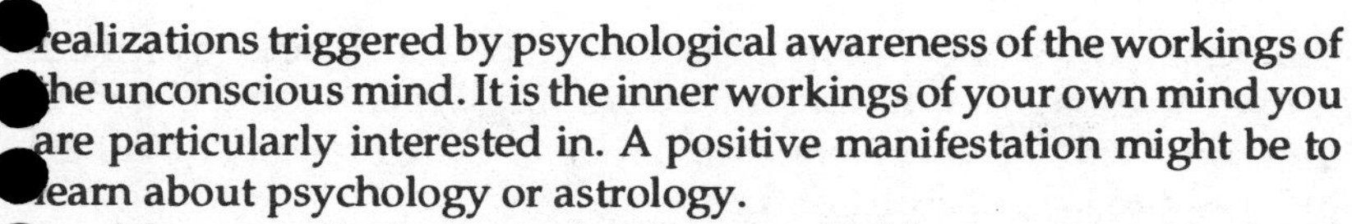

realizations triggered by psychological awareness of the workings of the unconscious mind. It is the inner workings of your own mind you are particularly interested in. A positive manifestation might be to learn about psychology or astrology.

More negatively, you might be surprised by your own irrational thinking and sudden outbursts of emotion. Your feelings and attitudes will tend to be unpredictable and erratic. Positions change from day to day. You are probably reacting to unconscious impulses rather than responding rationally. Be aware of the people and situations which trigger your upsetting emotions. Others may be actively trying to manipulate you. Dismantle the triggering mechanism by understanding the pattern of manipulation and the complexes or fears in you which foster the responses. Gain a perspective on your emotional depths.

URANUS IN THE 9TH HOUSE

Uranus in the 9th house is a sign of radical changes in beliefs, either because you recognize discrepancies in your own thoughts or because you are directly challenged by others and situations you are involved in. You may have recently received new information which specifically contradicts your former beliefs, making them obsolete. Sometimes there is a rude awakening followed by a sharp transition; at other times, a slow transition may result from a constant series of small challenges. You can no longer depend on old axioms.

Mundane as well as religious, spiritual and philosophical beliefs are now in a state of evolution or turmoil. The 9th house is not only the house of all higher thoughts, but also the house of beliefs about yourself, your abilities and other people. You must reassess your position on a number of issues and be ready to make adjustments accordingly. You must also be ready to defend your position against the challenges of others. Disputes over philosophical differences are possible and one person in particular may be an agitator.

Consistent with philosophical differences are cultural distinctions. An awareness of practices in other cultures may be necessary if you plan to deal regularly with foreigners during the coming year. (Examples: a businessman dependent on foreign exchange, support or commodities; a foreign exchange student; or one who plans to marry someone of a different ethnic background.) Ignorance is not bliss in this case and can lead to conflict if you do not make the effort to accommodate the beliefs and customs of others. In very negative manifestations, you may not be able to weather the cultural distinctions easily. Ignoring the differences in customs may make it difficult for you to establish trust and understanding. Practices and beliefs in your own culture or religion will not be accepted readily by others and need to be explained or defended.

Quick trips and hectic travel are more likely than leisure pleasure cruises. If you have dealings in foreign countries, you generally travel for business reasons. The trips are more commonly stressful, especially if you are having problems with your foreign counterparts. If you travel for pleasure, chances are you try to pack too much into one trip, in which case travel can become more nerve-wracking than fun.

Legal problems or dealings with lawyers are possible during the year, but this interpretation is not the norm and occurs only occasionally. Understanding the legal system is like dealing with a foreign culture and generally involves learning to cope with the peculiarities of justice in America.

If you are in school, it is likely that your education will be disrupted by distractions or unexpected events. The college or university you are attending may not be offering the course you want or need to graduate. Your time may be split between two different campuses, or between employment and school. Your study hours may be repeatedly interrupted and your concentration broken by situations in the classroom or at home. It is difficult to think in a stressful environment. What you are learning may make you anxious if it contradicts your existing beliefs. If you are teaching others, your schedule may be changed or broken up. Changes (e.g., renovations) occurring in the classroom or home force you to shift from one place to another or suspend teaching for a period of time. This is a good time to be self-taught. Independent study may prove more rewarding than a structured learning environment. You may elect to study on your own or take a correspondence course during the year.

URANUS IN THE 10TH HOUSE

Uranus in the 10th house shows that professional changes are likely to occur. In most instances these changes are major and involve a switch from one career to another or from company-oriented employment to self-employment. You will tend to be restless during the year. Professional freedom is usually an issue and you will not submit quietly to authority figures, especially if they are unpredictable or if the main emphasis seems to be on restriction of goals and frustration of success. If this is your situation, you will undoubtedly rock the boat.

You need to function as independently as possible for your creative urges to flow. If you are unhappy with your boss, transfer to another department or office location. If you are unhappy with your present job, find a new one. If you are unhappy with your present profession, jump careers entirely and start out in a totally new field, especially if Saturn is also in the 10th house. Consider becoming self-employed since you have the need and ability to function independently. Breaks in employment usually mark the period of transition. You may decide to stop work entirely or retire. There is an outside chance you may be fired unexpectedly from your present job if your performance is inadequate or if the company undergoes reorganization/merger.

Those who do not make major career changes might feel restless at work, and easily bored with repetitive tasks or distracted by disruptions. Use this time to integrate new ideas and systems into your daily procedures. You need a variety of tasks or a change of pace. There are many ways to incorporate change into the office routine and there is always room for improvement. You might become aware of numerous and continuing daily disruptions which prevent you from functioning at your best. Distractions can draw you away from your true purpose or job description. The entire office may need reorganization and management will usually welcome constructive criticism and enlightened input. Don't be afraid to make suggestions meant to streamline office procedures for greater efficiency. Extenuating circumstances may make it difficult for you to plan out your day. The most common situation along this line is business relocation or renovation. If your office is preparing for a move, setting up shop in new quarters or rearranging old ones, it may be tough to adhere to a schedule and plan out your days in advance. Another possibility is an office agitator who regularly disrupts those trying to do their job.

If you are not working at this time, you can still make major changes, usually in your life direction. Decisions may not be made quickly and easily and the tendency is to be erratic. Major lifestyle changes may be considered including divorce, separation, or major relocations, possibly overseas. You may move away from your parents or they may relocate to a retirement community in the "Sun Belt." If you are still living at home with your folks, you may disagree with their authority over you, especially if you are of age and need greater freedom. Demonstrating your maturity is the quickest road to independence.

URANUS IN THE 11TH HOUSE

Uranus in the 11th indicates future goals are being reassessed. B the end of the solar return cycle, you may end up doing somethin totally different from what you had anticipated earlier in the yea Changes are not always initiated by others, but many times this is th case. Unexpected situations involving others may encourage you t change your direction. For example, if your spouse loses his or her jo during the year and decides to becomes self-employed, you ma agree to be his or her business partner. If your best friend chooses to move to California, you may want to go along. His or her person change provides you with an opportunity for a new future directio For this reason goals are somewhat unpredictable and subject to radical revision. Changes and uncertainties originating with othe commonly affect you directly, encouraging you to modify your ow goals accordingly. The new goals you formulate are a major departure from the norm or past expectations and dreams. Althoug sometimes triggered by others, new goals are expressly your own an representative of individuated needs. This is the time to capture the essence of your abilities and desires in a direction tailor-made for yo alone. Though you may ride on the coattails of others at first, yo must set your own path. Goals may change drastically early in the year, or evolve continuously over the solar return time period. Eith way, you may have several false starts before full implementatio This is not a year to be very fixed about your future plans since as you acquire new information, you will continue to make adjustments.

Friendships may come and go during the year, old ones fading new ones arise. Some changes will occur slowly, resulting from a long period of gradual distancing marked by a decrease in commo interests. Other changes will occur quickly, the product of sudd attractions and/or separations. As old friendships end, new ones wil take their place. There are several common reasons for this transitio You, yourself, or a good friend may relocate to another city or sta If there is not a physical relocation, sometimes there is a mental relocation, one in which your interest shifts from one area to anothe leading you to join new groups and meet new friends while spendi less time with old ones. Only in rare instances might you have a falling-out with one particular good friend. This is a very negative manifestation and during this period of time your friend may chan greatly, appearing erratic and unpredictable.

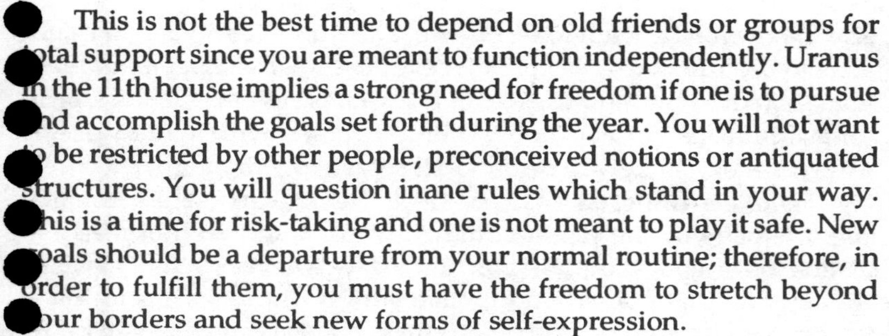

This is not the best time to depend on old friends or groups for total support since you are meant to function independently. Uranus in the 11th house implies a strong need for freedom if one is to pursue and accomplish the goals set forth during the year. You will not want to be restricted by other people, preconceived notions or antiquated structures. You will question inane rules which stand in your way. This is a time for risk-taking and one is not meant to play it safe. New goals should be a departure from your normal routine; therefore, in order to fulfill them, you must have the freedom to stretch beyond your borders and seek new forms of self-expression.

Peer pressure is noticeable with this placement and can cause a dilemma for those caught in a transitional period. Many fear what others will think if they break away from the pack and become their own persons. There is a push-pull sensation associated with any relationship, group, or friendship as one wishes to belong (become the same as), while at the same time recognizing the drive to be different. This is why it is helpful for some friendships to change. Maintaining the old environment while striving for new directions can cause great anxiety. The transformation of the social milieu generally allows one support for new endeavors and risk-taking while still lending a sense of belonging.

URANUS IN THE 12TH HOUSE

Uranus in the 12th house is the sign of the closet personality. During the year, your thoughts may be wild, but your outward demeanor remains conservative. Others will be unaware of what is really going on inside you, and usually something is going on. You have the ability to hide things. Private matters, internal thoughts or choices will not become public knowledge. A dichotomy exists between what you truly do or think, and what others are actually aware of. Distinctions may exist solely on the mental level or may also manifest into the physical realm. This is a time when you will tend to take liberties without anyone being aware of what is going on.

Generally, the fear of disapproval is common with this placement, and for one reason or another, you feel you must be on your guard. In all likelihood, you are involved in a new situation which is very different from what is normally expected of you. If you are not free to express yourself naturally and openly, you will do so secretively. Everything will tend to stay quiet unless Uranus transits across the solar return Ascendant during the year. If this occurs, you or someone else may wish the truth to be known. Difficult situations you might want to keep quiet usually involve sex-related activities such as initial sexual experiences, homosexual or bisexual preferences, affairs, birth control practices, or abortions. With this placement, matters of this nature are more likely to be handled secretively than openly.

Positive manifestations include the discovery of an unusual talent which you are just becoming aware of and are not ready to demonstrate. For example, if you find that you have psychic or artistic ability, you might be afraid to make your talent known before it is further developed. You choose to remain quiet about what you know or can do until a later date when you are ready to subject yourself to outside criticism.

Secrecy may lead to anxiety and nervousness over discovery. Furthermore, already existing negative feelings or fears will be made worse by the pressure of concealment. Thoughts and actions are such a break from your normal pattern that they appear to be irrational. Dilemmas cause you to flip-flop on important issues without understanding why. If you experience extreme psychological discomfort, see a counselor. Trying to survive in a vacuum can be frightening. Talking openly with a safe and understanding therapist will help you to assess objectively your situation and the choices available.

On a positive level, you develop a deep faith in your own individuality, learning to trust your unique inner qualities which free you from the limitations of conformity.

CHAPTER TEN

NEPTUNE IN THE SOLAR RETURN CHART

Introduction

Neptune is a slow-moving outer planet and therefore travels with the Sun, moving three houses clockwise around the solar return chart from year to year. It slowly shifts from an orientation through the angular houses, to passage through the succedent houses, and then through the cadent houses. Finally, Neptune repeats the cycle again by slipping into the angular houses once more.

Neptune is many things on many levels, but first and foremost it is the aspiration for a higher manifestation as opposed to a lower one. The distinctions between higher and lower influences are not limited to the spiritual plane alone, but can occur on the emotional, mental and physical levels also. At all times, it is up to the individual to steer his or her consciousness towards the highest manifestation possible. For example, Neptune on the spiritual level is associated with Universal Oneness, Karmic Laws, Ideals, God, and higher beliefs which form the backbone of spirituality. A strong sense of trust in God and the Universe supports the growth process on the spiritual level, but has ramifications on the other levels as well. Each level of experience supports and triggers the others. It is all the insights on all of the levels which eventually help us towards a more rewarding and fulfilling lifestyle. The lower manifestation of Neptune on the spiritual level is disillusionment with higher principles. Here the individual falls off the true path and becomes entangled in fanatical beliefs or outright fantasy. Spiritual despair rather than enlightenment is the result, and the support needed for growth on all levels is thwarted.

The same dichotomy of higher and lower is present on the emotional, mental and physical planes also. At the emotional level, the individual is capable of great compassion and sensitivity to others. This is a time when empathic understanding strengthens the bonds between loved ones. The lower manifestation of Neptune is a susceptibility to anxiety and worry. The sensitivity which is meant to foster true understanding instead heightens a sense of vulnerability to life, others and the future. There is no trust in God emanating from the spiritual level to support decisions. In negative situations, true understanding of others does not develop because the individual is too involved with personal feelings and issues to focus on others.

At the mental level, creativity and inspiration help to expand the individual's intellectual capacity. Neptune is more closely associated with the right brain than the left, and indicates the ability to let the mind float free to new and better insights. The lower manifestation is confusion and deception. What you are told is different from what you intuitively feel. Thoughts contradict insights and you are unable to comprehend the truth. There is no trust of the higher right brain inspirational or spiritual processes. Creativity is thwarted. Without an understanding of the big picture, mental energy is wasted through a lack of cohesiveness as one begins to focus on the insignificant, confusing details.

And finally, Neptune on the physical level is service to others. Principles which have filtered down from the various levels and have been understood in each of the higher manifestations begin to flow into daily practices on the mundane plane. There is a consistency: as above, so below. It is possible to physically manifest the spirituality to which you aspire. When only lower manifestations have filtered down to the physical level, confusion, disorganization and exhaustion are most apt to occur. You lack a total concept necessary to unify your actions and prioritize tasks according to their importance. Neptune at its highest level of manifestation on the physical plane is a direct reflection of the enlightened promise made at the spiritual level.

In all of the houses, either the higher or lower manifestations will be operating. The more the individual focuses on the higher energies, the more cohesive and insightful actions will become. The more the individual is sensitive to spiritual insight, the easier it will be to deal with issues on every plane. The unifying principles of the spiritual level and the higher manifestations combine to focus energy and understanding right on down the line.

Planets Aspecting Neptune

For Neptune's aspects to the personal planets, refer to the earlier chapters in this book. For example, Neptune aspects to Venus would be given in Chapter 5, Venus in the Solar Return Chart.

Neptune in the Solar Return Houses

NEPTUNE IN THE 1ST HOUSE

In its very highest manifestation, Neptune in the 1st house is associated with a renewed commitment of the self to a higher force or principle. This placement implies spirituality in being and in action. It is not enough to believe in higher forces—you must also act accordingly, practicing what you preach. During the year, the individual is seen as a part of the Universal Whole, possessing both personal insignificance and great spiritual importance at the same time. Insignificance occurs when the individual tries to stand alone. Self-centered needs and egotism are dwarfed in comparison to the spiritual process, and do not matter in the long run. If the will is not in keeping with the Divine purpose, it is limited and meaningless. Purely personal endeavors tend to get sidetracked before completion since the needs of others are more pressing, understandable, and important within the context of the Whole. For this reason, one tends to be egoless at this time, giving unselfishly to others and subjugating the self to a higher purpose.

Even though this is a time of personal insignificance, it is also a time of great spiritual importance for the individual. If you choose to view yourself and your soul within the realm of Universal consciousness, you will become aware of the role you play in the spiritual process and your ability to affect another soul's growth and potential. This is a time when you might understand some truth others miss, and if you are willing to give unselfishly, you can be a beacon for those who are blind. Situations may be subtle or blatant, commitments passing or definitive. In less obvious interchanges, compassion is increased, and you naturally want to help others. You become the Good Samaritan giving your time and energy to those in need whether family, neighbors or friends. In blatant interchanges, there is a definite request followed by a firm commitment to a cause, task or individual. You know what you are getting into. Some will chose to help many; others, only one particular person. As a rule, the process of assisting others comes easily and naturally when there is an awareness of a principle higher than one's own ego. During these times of giving, you will be encouraged to grow spiritually. Tasks usually exceed previous understanding and capabilities, and you must rise to the occasion through further enlightenment.

Certain cautions are necessary. Be sure the person or people you assist respond to your help in one way or another. Either their condition should improve, or they should learn new patterns of behavior, possibly responding with love and kindness towards you or someone else. It is not essential that you personally receive a return. This placement is associated with selflessness and what you are doing may specifically require an unselfish act of love. However, don't repeatedly waste your energy when there is no improvement at all. There are exceptions to this rule. Certain conditions are degenerative by their very nature, and do not lend themselves to improvement. For example, you may be caring for a person with a debilitating or progressive illness.

The manifestation of giving is meant to be helpful to the receiver, not draining for the giver. This is a year when you may be drawn to people in constant need, and in very negative interchanges when your help is not accepted, appreciated or effective, you will be drained. Do not become victimized by alcoholics, drug addicts or those who really require professional help, but refuse to seek it. Do not become so enamored of another person's problems that you close your eyes to your own welfare. It is possible to enter or remain in a situation which is personally difficult for you and has no easy ending or way out. You adapt so easily to those around you that you might adapt yourself into a psychologically unhealthy situation. Maintain a sense of spiritual purpose. Regardless of your circumstances, whether you are involved in a positive or negative gift of love, push for insight and growth.

While you are focusing on higher thoughts, you may have more difficulty with personal direction. You may not always understand your own actions, and can feel confused, indecisive, or lacking in ego definition. Surefootedness and stubborn opinions are unlikely because you are easily swayed. Self-knowledge will be elusive as you continually discover previously unknown facets of your personality. Preferences and abilities are in a state of flux. Unfortunately, it may not be easy to relate to your own quirks, or see your own contribution to problems and issues. This is especially true if one is not working towards spiritual enlightenment. Instead, you can be evasive or deceiving, lying when pressured. If the future appears uncertain, this is a time to go with the flow of events even though you are not sure where it will lead. Perhaps you are choosing to wait for another person to make a move or decision and this contributes to your uncertainty.

In rare instances, drugs, alcohol or personality aberrations can be the cause of your confusion or weakened vitality. Ongoing addictions can be particularly detrimental this year since this is a time when your body would tend to be overly sensitive to all medications, even prescription drugs. For this reason, alternative medical treatments might be more beneficial.

When living on the higher Neptune level, confusion, indecision, and drug-related problems are unlikely. You are much more likely to act on intuition, instinctively flowing with the Higher Self.

NEPTUNE IN THE 2ND HOUSE

Neptune in the 2nd house generally indicates monetary uncertainty lasting for most of the year. Those who are self-employed or working on a commission basis already have incomes which fluctuate. They can never be too sure exactly how much money will be earned. This is a common occurrence for these professionals, but there may be reasons for even more variation than usual at this time. One might be opening a new shop or pushing a new product, and a degree of uncertainty surrounds the hoped for yet unproven success. For those who are on a fixed salary, this placement may appear more threatening. Supposedly, you already earn a specified amount. If this is your situation, a number of possibilities exist. Rumors of layoffs, pay cuts or salary increases will make the exact amount of your earnings appear uncertain. Pay cuts may never materialize, while promised raises can be put off indefinitely. Rumors might be all that exist for the entire year. Salaried positions which are funded by grants will not be confirmed until the money arrives. Part-time employees with flexible hours have changeable incomes, and might only report when work is available. Seasonal fluctuations affecting some professions dictate the number of hours worked.

The salary situation is usually 99% uncertainty with no actual loss of employment or funds, but the situation can make you anxious. Decisions you make during the year might also lead to fiscal uncertainty. A leave of absence without pay is an option you could choose if you are pregnant, desire more time with your children, or plan to travel or study. Long-term projects without any guarantee of financial return are another option. You may volunteer your services. These are just some of the possibilities indicated by this placement which make money seem uncertain.

Financial anxiety is a problem especially if your income appears to be totally unpredictable or undependable. If you are presently unemployed, but share resources with another person, you may worry about his or her income as well as your own lack of funds. This is a time when it might be essential that you cope with some financial uncertainty in order to progress. However, it is not essential that you become anxious. That is a conscious or unconscious choice. For example, aspects to Neptune in the 2nd house from the 11th can indicate that goals are initially or basically economically unrewarding, or can only be accomplished with a degree of financial risk and uncertainty. If you can cope with the uncertainty, you may be very successful later on.

Some individuals have an unstructured attitude towards money and will not be very concerned by fluctuations. For others, uncertainty may be a product of the accounting system or lack thereof, rather than related to actual income. Hand-to-mouth economics prevails over budgetary restraints. Strict savings programs are unlikely at this time. The tendency is to be very free with money and not tightfisted. What money you have can be drained away unnoticed or used to help others. You easily give away extra funds or sacrifice your own financial security to assist those in need. The lack of concern may or may not be to your advantage. If you are too busy to be unduly concerned with money, this is no problem; however, if your present lack of financial accountability limits your progress or threatens your security, it is time to handle the situation in a more structured way.

The act of giving should be an act of love and not a product of low self-esteem. You must reassess issues of self-worth if you consistently remain in situations where you are unappreciated. Self-deprivation is not to your advantage and is not indicative of a higher consciousness. It is a manifestation associated with an inability to value oneself. If you remain in a low-paying job because you doubt your abilities, or a high-salaried, unfulfilling position because you are only worth the amount of money you bring in, you suffer from low self-worth. If you truly valued yourself and your abilities, and understood your spiritual purpose, you would move on. It is important to comprehend the mundane symbolism associated with finances. Monetary problems are meant to signify issues in personal assessment and appreciation. Issues in these two areas are interconnected and difficulties should be addressed.

During the year, money and possessions could be seen as unimportant in the long run. Other priorities tend to be stressed more than a good cash flow. Values are fluctuating towards a nonmaterialistic emphasis, and inner qualities and spiritual ideals gain significance as possessions or money lose ground. The appreciation of inner beauty increases over anything of external value. Personal satisfaction and a sense of fulfillment are now more meaningful, or at least they should be. The plane of matter needs to become consistent with spiritual purpose, and you must start to use possessions and money in a more enlightened manner. A simple desire for or dependency on funds will not suffice. Fiscal fluctuations will produce great anxiety in those fixated on their wallets. Use this year to put material possessions in perspective. Your present priorities set the stage for the future. Do you value cash more than health? More than job fulfillment? Could you use funds in a more meaningful and helpful manner? Do you keep yourself in a constant state of anxiety over money? This is a time when one is asked to trust the financial flow. You can live on less, take risks and still survive, going on to new opportunities. Those who are able to trust will experience minimal discomfort with this placement; those who are unable to do so will experience great anxiety. If you believe you deserve to be prosperous, you can simply trust in the flow of goodness and bounty while moving towards the fulfillment of a Universal need. Sustenance flows to those who work for the good of all.

NEPTUNE IN THE 3RD HOUSE

The most positive manifestation for this placement is an ability to be open to new information and new ways of looking at life without prejudice or judgment. Saturn is equated with reality structures which help us to categorize information quickly. Our criteria are set.

But sometimes structures are more limiting than helpful, and need to be broken down so new realizations can arise. If we continue to pattern information in the same way, we will never realize totally new forms of organization. The thought pattern associated with Neptune is very unstructured, general rather than specific, and for periods of time resembles no pattern at all. There are no lasting mundane criteria by which to assess individual pieces of information. The emphasis is more on the big picture and a total reorganization from the established structure. During the changeover, you may not be sure what is true. While Neptune is in the 3rd house, you are asked to acquire information without judgment, at least for the time being. Explore concepts without a preconceived notion of what you are looking for or what you will find. This openness is needed for major realizations to occur, and new ideas or perspectives can arise which you have not and would not have previously considered.

Neptune is also associated with higher forms of thought and profound questions that cannot be answered easily. There is the tendency to be very concerned with spiritual issues and the practical applications of higher concepts to the mundane level. A new sense of spiritual purpose could influence your daily activities. Intuitive insights are common, and often the shift in your understanding of reality is caused by the infusion of both emotional and spiritual information into conscious awareness. Do not box yourself in. Allow the mental transition to proceed at its own pace and time. You cannot force insights, nor can you hold them back. Flow with your feelings and realizations, not making concrete demands for the future. Some realizations may be beyond language and cannot be fully translated into words.

The negative side of this Neptune placement is the tendency to be easily confused and distracted. Your lack of emphasis on the "here and now" makes it more difficult to focus on practical matters. Your mind is becoming more aware of subtleties. You no longer have to be confronted with physical evidence to sense what is real and true. Intuition is strong and you are open to knowledge through the alternate channels of intuition and insight. While your mind is expanding rapidly on a spiritual level, it loses some of its desire to concentrate on maintaining daily patterns and physical order. This new sensitivity tends to overload your senses with subtle information, making it more difficult to deal with and remember mundane details. Right-brain insights predominate over left-brain concentraon, and distractions can occur with or without any perceived trigger. Your thoughts will be pulled away repeatedly from the task at hand to contemplate some new and sometimes vague concept or fantasy. Remembering to pick up the clothes at the cleaner's does not seem so important when you are realizing your relationship to the Universal plan or fantasizing about some new love and the future. Forgetfulness is common, or to put it more definitively, you do not care to be unduly concerned with details in the mundane world. The shift also makes it more difficult to discern between real sensations and anxieties. Premonitions and fears appear the same and one has the tendency to worry. Misinterpretations of insights can occur until the powers of discrimination are enhanced. Because of the mental uncertainty caused by the new information and thought patterns developing, you might give mixed messages to others until your system adjusts. Therefore, verbal communications are subject to misunderstanding and you have to work consciously during this time to communicate more effectively.

Decisions are harder to make, especially when they involve major choices with limited information and no guarantees. Some individuals become immobilized by the decision-making process. They become very concerned with doing the right thing, wanting a guaranteed result when none can be given. A lack of conviction may cause you to expect others to make decisions for you. Numerous pieces of advice will only add to the confusion. Make tentative plans as you proceed and save room for adjustments as new information becomes available.

Your mind is very susceptible to alteration. This may be a time when you explore different mind-altering experiences from drugs to meditation. Be an educated consumer and know what you are getting into. If you are on any medication, learn the side effects of the drugs you are taking. A pattern of confusion and low vitality may relate to a prescription you are on. A very negative but rare manifestation is substance abuse. Metaphysical education is associated with the Neptune process and could help trigger the spiritual and intuitive insights needed for growth and awareness. Use this time to work with the finer experiences associated with life and perceptions.

NEPTUNE IN THE 4TH HOUSE

Uncertainty is connected with the home or your living quarters, and you may not be certain where you will be living in the future or who will be living with you. Issues involving relocation are common. Either you yourself or the person/s you are living with might be considering a job-related move. In all likelihood, the exact destination or date of departure is not set. Also, the number of people going along might be questionable for a period of time. On the other hand, if you and your live-ins are happy with your employment situations, you could decide to look for a new residence in the area. Since there is no pressure to move immediately, you can look, but still feel unsure as to when or even whether you will move. House construction can be delayed indefinitely if you plan to build. Those who stay put might have family members coming or going during the year, causing confusion regarding who will be living with whom, and when. This is especially true if you have college-aged or recently divorced children or elderly parents. If any of these people are in a situation of need, they can come and go during the year without a lot of notification. These and other circumstances surrounding your living arrangements leave you with uncertain domestic plans during the coming year. It is most likely that major changes will not be a reality until the final three months of the solar return year.

You can have a family member requesting assistance at this time. Physical, financial, and emotional needs are possible. Issues are not necessarily serious; for example, you can help a relative build a new home or care for an infant. Use this time to share with family members. Compassion is increased, and one relative in particular could depend on you for support and encouragement. During the year, an older person might become forgetful, especially if he or she is at an age when senility is a problem. Medications can slow the thinking and physical vitality. Drug side effects should be considered if there is an appreciable change in his or her personality. If substance abuse is already occurring, confusion will be more noticeable at this time. Adults with elderly parents could consider having them move in either temporarily or permanently. Retirement communities, nursing homes or visiting nurses are other options. If you are caring for an older relative, indecisiveness and a wait-and-see attitude are common. It takes time to pick a course of action. In some way, either the health or mental capacity of a family member can directly affect you and any major decisions you must make. Uncertainty surrounding this family member leaves you without a clear personal direction or goal since you cannot be sure how much assistance he or she will need in the future.

Domestic uncertainty is usually coupled with a lack of emotional definition. Your feelings may not be clear to you, especially if the Moon is not strong in the solar return or has a conflicted interpretation. Emotional dilemmas can divide your feelings between two or more options. If you are dealing with a very withdrawn family member, you might not understand exactly what is going on. Misunderstandings and confusion persist; clear information is not available. For some, emotional clarity will not be important. The tendency is to offer assistance regardless of feelings. Emotions are generally on outflow, not intake, and the needs of others supersede your own. Self-sacrifice is possible if you have mixed feelings about someone who is in great need of assistance, but don't let situations become so lopsided or negative that they become detrimental to your own well-being.

Neptune implies increased spiritual attunement to your sense of purpose here on the earth plane by first washing away preconceived notions so that new sensitivities can be felt. Increased compassion in the home and for family members helps to establish a new code of behavior. The roots of this new code are in a spiritual identification with higher forces brought about by newfound sensitivity. The first manifestation of this new code is in the home, but spiritual insights defined now may be expanded to other areas of life as Neptune moves on.

NEPTUNE IN THE 5TH HOUSE

Spiritual concepts of unity among all human beings can soften the way you express yourself. A gentler involvement with the world is warranted at this time. As with all Neptune placements, higher principles are seeking a practical application in the real world. An increased sensitivity to specific situations, and uncertainty or confusion associated with actions or thoughts, tend to signal this infusion. While Neptune is in the solar return 5th house, spiritual concepts which are already understood need to find a mode of expression. Principles of Universal love can be expanded upon through compassionate interactions with lovers, children and others. Working with creative and artistic projects can also heighten your awareness. What is needed is an evolution in self-expression towards a more spiritual and insightful manifestation. Inconsistencies between the self and the Higher Self must be eliminated. You cannot believe one thing spiritually, yet express something else to others. The self (external expression) must be one with the Higher Self (spiritual ideal).

As you gentle your approach to the world, you may not be sure what you wish to express or represent. Tentative self-expression is common during the transition period. Sometimes external changes in the social milieu trigger the shift, but often gentleness is brought on by an increased awareness of the way you interact. Careful consideration is warranted. How are you perceived by others? Do you reflect your spiritual beliefs? Do you allow your true self to come through, or do you hide behind a persona? Is someone questioning the way you express yourself or what you purport to represent?

Consistency is important. Ask yourself, "Am I real? Or do blockages keep me from expressing who I really am, or what I really feel?" During this time, others tend to point out your inconsistencies. Occasionally, they add to your tentativeness by giving you very negative messages which are not truly insightful. Self-discernment is a task associated with this placement and you must learn to discriminate between helpful insights and negative comments. The lack of certainty along the way allows for the development of the softer side of your personality by making you hesitant enough to reflect on the other person's perception of you and what you truly wish to impart. For this reason, it's important to view self-expression from the other side, seeing yourself as others see you. Confusion about your identity is likely to continue until you begin to manifest the spiritual directive you purport to understand.

Increased compassion and sensitivity to others will bring about positive changes in self-expression. Loving and nurturing relationships can lead to an evolution in consciousness. Nonsexual spiritual relationships are possible with this placement, though sexual involvement is not necessarily prohibited. Nonetheless, it is the spiritual love which is important. You can love someone dearly without any thought of return or commitment. Assurances will not be necessary, nor will they be forthcoming. A lack of definition tends to permeate love affairs while Neptune is in the solar return 5th. You can never be sure where the relationship is headed, and the future is often left hanging. Sometimes you are not even sure when you will see each other again. This is a time to let relationships simply "be" what they are meant to be. Real limitations may or may not exist. For example, your lover might be away much of the time or previously committed to someone else.

On the other hand, the lack of definition or commitment could be totally confusing to you. You may not understand your lover's reservations and a sense of vagueness or mystery could surround your interaction. Conversely, you might be the one to want to keep things loose. In any case, guarantees are not given. Part of the problem might be your tendency to idealize the person or romance you are involved with. If your judgment is clouded, lovers can appear more attractive, attentive, or spiritual than they really are. Eventually these misconceptions lead to disillusionment. In the most negative situations, you are attracted to someone because one of you needs help and expects to be saved. Savior-victim relationships tend to be one-sided, with one person giving all while the other wants to receive.

If you have children, or deal with them daily, they can be a great source of insight and calmness. On the other hand, they might depend on you for direction, appearing lost when left to their own devices. This can be a time when children seem slightly out of character, acting up more than usual, especially at school. Or they might be extremely quiet. In either case, your extra sensitivity should be used to draw them out. Compassionate interchanges can help you to better understand them, their needs and their abilities. Spending quiet times with children can be mutually beneficial. This is a time when parents traditionally tend to sacrifice their own needs for the special needs of their children.

Increased sensitivity can also be channelled through creative or artistic projects, if you are so inclined. You might become aware of new subtleties in self-expression and design. New methods will focus on more delicate techniques. Aesthetic appreciation and insight also increase.

NEPTUNE IN THE 6TH HOUSE

Neptune in the 6th can mean working at a job when your heart's not in it, or when you really have no sense of direction or purpose. Most commonly, you are in a period of transition between jobs. You are in the process of finishing up an old job, business or career, and moving towards a new professional endeavor. Usually there is an overlapping period when you are not physically through with your former job even though you are mentally finished. Mentally, you are now involved with your new position even though the physical transition has not yet begun or been completed. So you are just marking time, picking up a pay check, or existing in limbo until everything is set for the transition to be accomplished. It is unlikely that this will occur until the end of the solar return year, or until

Neptune transits out of the 6th house, unless your new position is as probationary or "iffy" as the transitional period itself. Becoming self-employed generally falls in this category. A need for job fulfillment pushes you to find a more suitable position. The job you are leaving might fall short of your abilities. You may not feel as helpful or effective as you could be if given more opportunity. Misunderstandings between you and your employer are possible. A few individuals might be unemployed or drifting from one job to the next during this time.

Assuming you are happy with your present employment situation, this can be a time of job confusion or uncertainty. New office procedures might turn your workday upside down if inadequate training leaves you unsure of what to do. Job security in your present position can seem precarious. Rumors of layoffs, mergers, relocations, or shift changes are possible. Poorly defined threats to to your job are likely, but generally there is no loss of position or employment time. Coming reorganization plans leave future responsibilities defined and you might be unsure where you will be in one year's time. Or you could be involved in a project which has an indeterminate outcome. You and the others could risk your time and effort on a long shot project you are not sure will pan out. If you are the owner of a business, you may be considering a merger, buyout, partnership or stock option. You will find it difficult to make concrete business plans for the future until certain issues are settled. Misunderstandings with others can occur during the year regardless of your level of employment. It is to your advantage to communicate directly with business associates. Do not rely on intermediaries to convey your messages.

Illnesses and diseases, if they occur, tend to be more difficult to diagnose while Neptune is in the 6th house. Generally you are the person seeking diagnosis, but if you are responsible for a child or an elderly adult during this year, it is possible that the vague and undiagnosed health problem is his or hers rather than yours. The health provider may be baffled by the symptoms and recommend several different courses of treatment. He or she may not be sure exactly what will work. The tendency is to try one method first, and then move on to another form of treatment if needed. A second opinion might be advantageous. You can be successfully treated without ever really knowing what you had. The origin of an illness needing diagnosis and treatment may not be clearly understood for most of the year, or until Neptune transits out of the 6th house. Symptoms are generally vague, conflicting or intermittent. Complaints of tiredness are common. A case of the flu might hang on longer than expected, or leave you feeling less than "right." Allergies to smoke, aerosols, pollutants, medications, and foods are more likely to affect your lifestyle or diet during the year.

The increased sensitivity associated with Neptune is apparent in the physical body's acceptance or rejection of certain daily work, living or eating habits. Emotional upheavals and stress are more apt to affect your physical form and aggravate existing illnesses, causing symptoms to reappear or multiply. If stress is coming from your job, realize that health is more important than work. Spiritual beliefs and attitudes need to become a part of your daily routine while certain detrimental habits fall away. If you cannot maintain a spiritual perspective during your daily routine, it may be time to change the routine. Cultivate serenity.

NEPTUNE IN THE 7TH HOUSE

Relationships tend to lack definition and certainty during the year that Neptune is in the 7th house. There is a vagueness about involvements which manifests in a variety of ways. This is generally not the time to expect firm commitments. Relationships, whether good or bad, tend to seek their own level of involvement and are not subject to categorization. It is unlikely that you will get married during the year, though you might discuss the possibility. The tendency, though, is toward all talk and no action. Strong placements in the 5th house, aspecting Neptune in the 7th, tend to indicate that the involvement is more closely akin to an affair than an established partnership. Perhaps you are not quite sure where the relationship is headed. This can be true even though you have been together for a number of years. Your partner may be contemplating career or personal changes which will affect you either directly or indirectly. You might be asked to make a leap of faith, trusting your partner's instincts and providing the support necessary for the transition he or she wishes to make. Generally, you must do this without any guarantees of success or return. Occasionally, you will be asked to sacrifice your own needs for those of your partner.

Relationships can be downright confusing and anxiety-producing with little sense of direction or destination. It is best just to let situations evolve and go with the flow, not making concrete plans for the future. Relationships may grow or they may die, but they cannot be planned out or structured this year. In the most negative manifestations, you waste your time trying to rescue someone when the individual really does not want to be saved or even understood. You try to force your perception of the future onto someone else's right to make choices. If you continue to push against all reason, you will lose your sense of self, and eventually become victimized by your own misguided efforts.

Misunderstandings will be a real danger. You must be more mindful of the need for clarity. Perhaps your partner says one thing and does another. Actions and words conflict; double messages are possible. Confusion ensues. Learn to collect information from more than one source. Do not rely solely on verbal information. You must trust your own judgment or you will be too easily swayed by others. It is possible that you will be lied to or deceived this year by an intimate partner, probably not a casual acquaintance. Idealization plays a major role in the deception process; in effect, you fool yourself as much as allowing yourself to be fooled by others. Use several means to assess reality and remain grounded. Always be true to yourself. There can be a fine line between deception and misunderstanding. Only time and developing circumstances will tell which is which. It is possible to contribute to your own difficulties without seeing yourself as a primary cause. Give careful consideration to your partner's side of the story.

On a more positive note, Neptune can show increased sensitivity within a relationship. This can be a time when you develop a spiritual union marked by greater caring and compassion. Understanding can be established without words. Spirituality grows if you both follow your intuition and draw on Universal connections. This is especially true if you are married to the person you care most about or have been able to establish an enduring relationship over a period of time. Seek a feeling of connectedness with another.

NEPTUNE IN THE 8TH HOUSE

The most notable mundane manifestation associated with this Neptune placement is confusion or uncertainty surrounding resources and finances you share with another person. A variety of situations can arise. Your partner or roommate may be laid-off, unemployed or too sick to work. If he or she has a steady income, other circumstances including wage freezes, delayed promotions, or salary negotiations might make the future income unclear. You may hear that more funds will be made available eventually, but wait much of the year for the money to materialize. If present earnings or holdings are very diverse or in the process of changing, your partner's income might be incalculable. If you yourself are dependent on parents or others for funds, money might be given to you only occasionally or as the need arises. There will not be any schedule of receipt or determined amount, so you cannot depend on a steady allowance. The giver may only give when he or she remembers and forgetfulness is a problem. If you had been receiving money in the past, the flow could stop without your understanding why. Uncertainty about money due you is also possible. If you are waiting for a suit, debt or insurance claim to be settled, the issue may go on for most of the year.

You might be totally dependent on someone else, or another person might be totally dependent on you. If you are the one totally dependent, your future financial situation may be very iffy since others tend to be undependable to begin with. If the other person is totally dependent on you, you can be taken advantage of, sacrificing your own financial security to assist someone else.

Giving away money to charitable organizations is a positive trait associated with this placement. You might wish to consider tithing some of your income to a church, synagogue or charity.

You could become more aware of subtleties in human nature. New information can come to you psychically or intuitively, but you are most apt to be aware of psychological idiosyncrasies. You can grow more tolerant of human frailties and other people's problems, or you can totally lack understanding and compassion. Psychological issues may be so confusing to you that you cope by withdrawing. Experiences might include coming into contact with those who suffer either physically or mentally. If you are considering volunteer work, try visiting the sick or working at a hospital.

Sexuality can be combined with spirituality. If you have been promiscuous in the past or generally indiscriminate, you might reconsider your criteria for involvement. Occasionally, questions of sexual orientation might arise, but the realization of the beauty associated with a totally loving sexual relationship is the most positive manifestation. An openness to tantric sexual practices might be advantageous. Explore your own sexuality and learn to listen to your own body. Foster sexual practices which are more consistent with spiritual endeavors and principles.

NEPTUNE IN THE 9TH HOUSE

This is traditionally known as the house of religious and philosophical beliefs. For those who are seeking to raise their consciousness through meditation, spiritual studies, prayer, or alternate realities, this can be a time of great enlightenment. A realization of God and the Christ-consciousness is possible. Mystical experiences occurring during the year could significantly change your understanding of yourself, the Universal Oneness and your purpose here on the earth plane. Strong realizations of this nature cannot be translated into words, and therefore may not be understood by others lacking the experience. This is a time when you move towards your own inner comprehension of God and the spirit of the law, while realizing that religion and the letter of the law are inferior attempts at definition and comprehension.

The danger with this placement is a tendency to be misguided. Because there is a movement from external standards (religion) to an internal realization (enlightenment), confusion and uncertainty may accompany the transition. You can be off track for a period of time and find it difficult to cope with philosophical and practical decisions. Mundane events and issues can test your new and old beliefs, pushing you towards further definition and understanding. Do not put your faith in a belief system which is totally unrealistic or impractical. Acute idealization is a problem, and unattainable expectations will inhibit your ability to function in the real world. If you push too hard for enlightenment, you will fall into the trap of focusing on the letter of the law while missing the spirit behind it; contradictions will arise. Others may be forcing you to pursue religious or philosophical systems which do not fit your needs or are inconsistent with your purpose. You are capable of being swayed by others.

The distinguishing criterion here seems to be one of understanding. If you can express a principle in words, but do not understand what you are saying, this principle is probably incorrect for you and your needs at this time. However, if you have captured the spirit of the principle within your understanding and know that it conforms to the Universal need for goodness, do not be alarmed by your inability to translate these insights into words. True understanding is, many times, beyond words. At the same time, principles and experiences which come through true enlightenment cannot be passed on to others who have not had the experience. Common ground for understanding will not exist. Grow to trust the inner process.

Others might be intolerant of your beliefs, or you of theirs. You must deal with your own misconceptions or with those of others. Occasionally you come in contact with fanatical beliefs, and in extreme cases victimization can occur. Harassment because of racial differences, sexual orientation, or ethnic and religious prejudice can occur. In extreme cases, the harassment leads to legal problems. Legal questions are likely to remain undecided for much of the year. If you are involved with foreigners, or if you are a foreigner in another land, cultural differences may lead to difficulties or intolerance. Misunderstandings are possible, especially if you do not understand each other's customs.

You may be thinking of attending school. If so, your course of study might be undecided. Perhaps the exact curriculum you need does not exist as a standard major, so you mix and match. You may be attending school only on a trial or probationary basis since you lack a clear understanding of your motives and goals. Financing for your education might be uncertain, leaving your continuing attendance up in the air. Another alternative is that you do not matriculate at all, but only consider the possibility all year long. But this is a good time to study religion, philosophy, or holistic concepts. This field of study can be particularly helpful.

NEPTUNE IN THE 10TH HOUSE

Neptune in the 10th house shows some uncertainty associated with professional situations and goals, and also nonemployment destiny choices. Your present job can seem unstable, especially if you hear rumors of mergers, layoffs, reorganization or relocation. A promised promotion or managerial change may be left hanging most of the year. The tendency is more toward uncertainty than actual problems, though occasionally individuals are unemployed or laid off for a period of time. More than likely, though, you will be considering some kind of career change, possibly leaving an old position or profession, future destination unknown. Even if you were to attempt to make concrete plans at this time, too many factors need to be considered and settled before you are guaranteed a future option, assuming guarantees exist at all. You must be prepared to adlib. This is a good year to handle career uncertainty and still go with the flow. The specter of the unknown should not stop you from making the journey. Begin even though the final destination is unclear.

The career changes you make this year are important and usually move you towards a greater sense of satisfaction and fulfillment in your profession. The spiritual qualities you believe in must find a stronger manifestation in your life through career pursuits. You will be looking for a job which fits your spiritual intent and emotional system. There can be a period of unemployment, because you will not want to settle for less. Professional self-confidence is essential to the process and you must believe in yourself and push for career options. If you cannot develop the clarity and confidence necessary to pursue what you really want, you will spend time rejecting jobs you do not want. Learning might be a positive assertive process, or a negative rejecting one. The end results can still be the same. You need to come to terms with the inner and the outer manifestation. Perhaps the inner need for fulfillment should be more appealing than money, status or recognition. Once you are over this hurdle, situations develop more steadily. Those who can sense the delicate inner desire for fulfillment and push for manifestation early in the year will move steadily towards a new professional endeavor.

Major destiny decisions can also be made during this year, or successfully avoided. Decisions can be so far-reaching that the individual refuses to even look at the issues or options available. Idealization of the present situation or future possibilities can block any assessment of reality. Lifestyle choices associated with marriage, divorce, or relocation are possible. Again, the emphasis should be on fulfillment. Changes in consciousness at this time push you toward paths which more accurately reflect your life purpose and spiritual beliefs.

NEPTUNE IN THE 11TH HOUSE

This is a time when you can develop a spiritual bond with others. You grow more sensitive to the needs and fears of friends and can respond in a way which is both helpful and insightful. Even if certain friends are mere acquaintances now, intuitive bonds can be formed with those who are equal to this kind of exchange. Relationships transcend self-centered needs and become manifestations of a spiritual connection. It is the inner qualities which make a true friendship. Compassion and empathy play a major role in your involvements. Those you are closest to could need your assistance. Use this year to develop a deeper understanding of what it means to be a friend. Master the art of giving as well as receiving. The purpose of this placement is to discover who your friends really are. Some friendships will grow stronger because they meet your spiritual impressions of what a friend can and should be. Others may fall away as you see through their deceptive facade.

Unfortunately, some friends are not as sensitive or developed as you. Compassion and empathy are sometimes one-sided. You could be sensitive to your friend's needs while your own needs go unmet. Your friend might want something specific from you without much intent of return. In extreme cases, you are asked to play savior to a victim role. Your counterpart needs your assistance, but at the same time his or her actions appear purposeless, self-destructive or contradictory. These savior-victim relationships are particularly disappointing. Remember that each person needs to contribute to his or her own salvation. Disillusionment is most likely to occur if you idealize your friend's potential without understanding present limitations. Confusing behavior and interactions might also exist, and if so, misunderstandings are likely. If there is no intuitive bond between you two, then what is left unsaid is uncertain. Try to clear up misunderstandings as soon as they occur, but realize that you and your friend might now lack a common basis for understanding, each of you needing to head in an entirely different direction.

All of the above information can also apply to group interactions. This can be a great time to join a spiritual study group to further your own enlightenment. But also realize that groups may not always live up to your expectations.

Early in the year, goals are not well formed. Even if you think they are, major adjustments are still likely. This placement is commonly associated with partial or uncertain goals. The tendency is to follow a whim without being sure of the final destination. Although you go with the flow, are you really sure where the flow is headed? Probably not! Sometimes the rug gets pulled out from under you and you are too busy with your immediate situation to sort out your future. For example, suppose you suddenly become unemployed or pregnant. Coping with your present situation will be your first priority before you start to reconsider your long-term goals. You cannot plan for the future until you settle things in the present. You can only anticipate the next step. Occasionally, an individual does not even realize a goal until the end of the year, but this should not stop you from working on it directly or indirectly for those 12 months. For example, if you return to school, you may be undecided as to your major. You have particular courses in mind, but you are not sure exactly how you will use the information professionally or what degree you wish to pursue. You are open to several different career possibilities. But you do take necessary courses all year and eventually the path becomes clearer. So reach for a goal even if you do not understand what your ultimate destination will be. Suspend judgment and proceed on a hunch. Many times the movement is toward a more spiritual manifestation. The quality of life is a major priority. This is a time when future plans grow more consistent with spiritual beliefs.

NEPTUNE IN THE 12TH HOUSE

Neptune in the 12th indicates the possibility of confusion at the unconscious level. In some cases, this confusion leads to free-floating anxiety. During the year, inconsistencies exist between what you experience intuitively and what you are told by others. Glaring contradictions make it difficult to know what to believe. There is a question here of who is right. Learn to trust your own instincts and Higher Self, while suspending judgment for the time being. There are subtle forces at work and the object of this placement is to become more sensitized to less apparent energies. You are aided in this endeavor by discrepancies between what you are told and what you sense. These inconsistencies force you to see two separate yet distinct pieces of information. If there were no contradiction, you would miss the more subtle message. Learn to be sensitive to what is left unsaid, yet do not confront others with your newfound truth. Discretion is needed. Also, do not become a detective sneaking around searching for confirmation. All will be made known to you. Time will eventually prove the information right or wrong. In the meantime, you can allow the two separate pieces of information to exist in your head for most of the year.

The inner self is not always right, though the Higher Self is; however, the Higher Self would never be preoccupied with the discovery of mundane proof. This is a sure sign that you are dealing with a fear rather than an intuitive insight. Anxiety is born of fear, while insight leads to enlightenment. Those who are obsessed with mundane truth must realize that neurosis is also associated with Neptune in the 12th house. Those who refuse to accept the truth they see, and consistently push for more proof, suffer great inner tension from suppressed information. Eventually, severe anxiety and even neurosis can ensue. True insight is used to develop understanding and compassion for others and is never used as an invasion of privacy. For this reason, the truth of the matter is not essential, only compassion and understanding are. These you can foster without all the facts.

Spiritual growth is very important at this time if you are to handle new information wisely. You need to develop a sense of trust in the Higher Self and God. It may be that you must subsist on trust for most of the year without any concrete proof or understanding until the year's end. The more you believe in yourself and the Higher forces, the easier this time of uncertainty will be. The most recent spiritual insight should be sufficient for the time being. Pushing immediately for further enlightenment leads to confusion and disillusionment. Wait until the time is right. While you are on the spiritual path, learn the value of not revealing everything you know or have experienced. Others may not be ready for inner truth and will be threatened by what you say. Unnecessary confrontations or challenges may ensue.

It is not enough to be materially successful; fulfillment is also important. Life situations tend to focus on inner satisfaction rather than immediate material attainment. Trusting in Divine guidance leads you to fulfilling circumstances. If you have a strong spiritual sense of what you want to do, trust the Universe to provide you with the means to accomplish the task. An act of service may be required to get there.

CHAPTER ELEVEN

PLUTO IN THE SOLAR RETURN CHART

Introduction

Pluto is a very slow moving outer planet and therefore, like the Sun, shifts three houses clockwise in the solar return chart from year to year. Also like the Sun, it shifts from angular placements into the succedent houses, then cadent, and eventually repeats the cycle anew.

In the solar return chart, the interpretation of Pluto includes an understanding of the issue of power and its various manifestations, which can be directed towards three different points of focus and through three different life processes. Complications are associated with all of these orientations and processes since each manifestation is multidimensional and complex, occurring not only on the physical level, but on the intellectual, emotional and spiritual levels as well. Within these complex situations, power can be focused on three different targets: the self, others, and/or circumstances. As a rule, the power to control oneself is mostly beneficial, the power to control others is mostly detrimental and the power to control situations varies according to the circumstances involved. By studying Pluto's house placement in the solar return chart, you can assess in which areas of life these control issues are most likely to arise. The scope and effects of the power issues are seen through the aspects to other solar return planets.

Power also manifests through three different processes. One may choose sameness, elimination or transformation. When you choose to keep things the way they are, it is implied that there is some force urging you to change and you are resisting this force. A possible power struggle might ensue. When you choose elimination, you relinquish control over some facet of your life or refuse to have further contact. When you choose change, you may be either yielding to an outside power or actively seeking transformation of that which already exists. Let us explore the power orientations and processes further.

Three Different Points of Focus

Generally, the focal point of **self** is beneficial. As long as you are using Plutonian insights and techniques to control yourself, to own your own power and to further your own growth, the process will be a positive learning experience, unless you overcontrol yourself, succumbing to hair-shirt asceticism, taking self-denial to absurd extremes. The greater your awareness of yourself and Universal principles, the more likely you are to make good decisions. You become the captain of your own destiny. Enlightenment leads to power and power leads to enlightenment as a rewarding cycle of manifestation is set in motion.

However, those who use psychological insights to control and block their own growth waste precious energy. They set up a negative pattern of stagnation wherein insights are not accepted as they are, but twisted to fit previously conceived notions about life and self. Rather than progressing through enlightenment, the individual uses insights as weapons against growth and understanding, opting for sameness despite new information. Contradictory realizations are not allowed to surface; instead they are either suppressed or misinterpreted. In very negative situations such as these, the power associated with increased Plutonian awareness and its creative potential is never realized.

The same thwarting of personal growth is evident when you shift your attention from self to the **need to control others**. By and large, this is not a good focal orientation. The symbolism of the planets exists in your consciousness so that you can grow and prosper from insight. Since growth begins at home, the main thrust of the focus should always be towards self, and it is generally unnecessary and also counterproductive to shift the emphasis to an external struggle. The most efficient use of power exists on the internal plane because it takes much less energy to control your own reactions than to seek power over someone else. Once power is removed from the inner self and used to control others, the effectiveness is reduced. Stalemated power struggles are particularly detrimental because they deplete resources. It is conceivable that there are life situations where it is essential to control another person. For example, children, disabled or elderly loved ones may not be able to make informed decisions. In these situations you might be asked or forced to wield power over another. You are meant to grow from all your experiences. Even in these instances, self-awareness and insight are crucial to the decisions you make. Start with yourself first, then work your way outward. Blaming others for your problems distorts the perception of self and the realization of your own personal involvement. Always be aware of the interactive process and the role you play in it. This internal perspective is of primary importance, much more so than the process of controlling someone else.

Trying to control life situations can lead to power struggles, but at some point in your life it may be essential to make a stand for the good of your own growth or that of world consciousness. Ultimately, it is much better for you to control yourself, but this is not always possible. Someone can be seeking to control you, your family or your livelihood. Certain injustices must be corrected for everyone to progress, and sometimes the only viable option is to work for change. Always be aware of the struggle you take on. It is usually easier to correct a situation than to convince your opponent he or she is wrong. It is easier to get forgiveness than permission. Go for the simplest task that gets the job done. For example, your office routine might be very inefficient and next to impossible to accomplish in a day's time. Perhaps you see where improvements can be made, but your immediate boss disagrees with you or thwarts your efforts. As long as you continue to try and convince your immediate boss of the need for change, you will get nowhere. If you can easily go over his or her head to a higher boss, this could be an effective way to handle the stalemate. Seek the quickest solution to a problem if one is available. Do not get locked into long, drawn-out struggles if they are not necessary.

Ultimately, power struggles can have either beneficial or detrimental results, depending on your perspective. In the external environment, you may or may not win, assuming there can be a victor.

More than likely, victory, if it comes, involves compromise for all involved. But the inner process of awakening one's own ability to effect a creative change in the environment is usually beneficial. You can make a difference. The creative potential associated with Pluto comes from a penetrating insight into the hidden worlds of the unconscious and an understanding of the Universal laws of Karma. Powerful insights should spring spontaneously from the situations associated with Pluto's placement in the solar return chart. Suddenly you can become aware of new forces, subtle and previously unnoticed, which seem to influence personal decisions and situations you are involved in. Once you have the insight, what you do with the information becomes crucial to the creative process.

The primary goal of the Plutonian process should be growth through awareness. Therefore it is essential, no matter what your focal orientation, to use information and insight gleaned from experiences to foster a new understanding of self and Universal principles. The main growth process begins and ends with the self.

Three Different Plutonian Processes

Plutonian power is associated with three different processes: keeping things the way they are, usually by resisting an internal or external force; changing the form of what already exists, i.e., transformation; or eliminating what is no longer useful or essential, a form of death. Any of these powers can be either beneficial or detrimental to the user or others involved. It is the purpose and intent of the process that is important.

The **power to keep things the way they are** can sometimes be an awakening to personal power. Implied here is a resistance to a force seeking change, and sometimes the intent of change is negative. The pressure can come from an external or internal source, and the changes desired may be either external or internal also. For example, if you are a recovering alcoholic facing a difficult life situation, you might feel the pressure to start drinking again. The pressure could come from your own wish to avoid major decisions, or from a person you are associating with.

In actuality the internal and external sources are really one and the same, a reflection of each other since internal needs draw external situations. In the situation given above, the individual seeks to remain a recovering alcoholic despite internal and external pressure.

This is a positive goal and if attained, implies personal power. It is also beneficial to remain the same when others seek to control your actions and thwart free will. Pluto represents the insight necessary to perceive manipulative efforts as they arise and avoid ploys whenever possible.

In some instances, the power to remain the same can be a negative manifestation, actually causing the stagnation of growth. If you resist all new internal and external insights which would enhance your growth or cause you to make changes, little progress will result. You will stall. When your purposes are not in keeping with Universal Good or personal benefit, you are more likely to be involved in a negative use of Plutonian power.

The process of **elimination** can be very cleansing, even though a symbolic death is involved. If you streamline your business or office procedures, letting go of compulsive and unnecessary activities, this is a death of sorts, but much to your benefit. So is a budgetary review which eliminates the fat from your spending practices and allows you to cut back on your work schedule. In more serious circumstances, perhaps all you can do is release a situation that is detrimental to your progress. Letting go can be a positive choice, but it might also be a negative one if you cut someone out of your life without resolving or releasing the conflict. If this is true in your case, avoiding persons or situations will consume your energy.

Transformation is the process whereby a situation, thing or level of consciousness is changed into a new, and hopefully higher, manifestation. Careers, marriages and self-images can all be transformed for the better. Commonly, though, we think of the transformation process in terms of psychological insight and level of consciousness. Information from subtle sources or the unconscious is gradually or suddenly made available to either the rational mind or the Higher Self, and subsequently a change occurs on the mental, emotional or spiritual level. Information that was previously unavailable becomes understandable through the process of insight, and power is released through new awareness. This can be a very enlightening and beneficial transition, but naturally it is what you do with the information that is important. If intellectual, emotional and spiritual growth are triggered by the new awareness, the process continues to be positive; however, if instead you use your newfound insight to control others or block growth, you have used your knowledge in a negative way.

Misuse of the Plutonian process creates an inability to continue on the life path until issues are resolved and lessons are learned. Stagnation occurs as the individual fails to progress to the next level of comprehension or misinterprets the task at hand. Power struggles can be the cause of such stagnation, and should be taken on only after careful consideration of the issues and implications. Power struggles are expensive in terms of time and energy. Some are essential to growth, others are self-inflicted. Know the source. Ideally, if you avoid a conflict in Pluto's solar return house placement, you should have more resources available for higher awareness. Conflicts consume energy and sap the strength necessary for the ascent. Without opposition, one can move quickly when concentrating on insight and awareness. Barriers can be torn down and the connections among all things become apparent.

But a world without struggle is the ideal, and not necessarily the reality. Your situation may require that growth be attained through struggle. The resistance may stimulate your awareness and creativity. Real progress might only occur after the resolution of conflict or during respites in the battle. However, within a long-lasting, stalemated struggle, one is more likely to get caught in a circular argument going nowhere. Nonproductive battles can be avoided by those who work towards increased awareness.

Planets Aspecting Pluto

For Pluto aspects to the personal planets, refer to the earlier chapters in this book. For example, Pluto in aspect to Venus would be given in Chapter 5, Venus in the Solar Return Chart.

Pluto in the Solar Return Houses

PLUTO IN THE 1ST HOUSE

Pluto in the 1st house indicates a strong desire for self-control and an emphasis on personal power issues during the coming year. Positively, you are capable of doing whatever you set your mind to. You have the ability to reshape your personality and control personal habits. Some individuals have used this time to quit smoking or drinking, or lose weight. This can be a great time for self-improvement since you have the ability to start a regimen and stick to it. The personal changes you are able to make may lead to career changes or successes in other areas either this year or the next. The power you use to gain control of yourself can ultimately be used to guide others to personal power in the years to come.

But a self-improvement kick is only the superficial interpretation for Pluto in the 1st house. Personal power in all situations is the real issue. As the year begins, you may not like yourself and may want to make sweeping changes towards controlling your own behavior and taking charge of your life. An awareness of personal power or the lack thereof becomes the focus for growth during the coming year. Often there is a question of who is in control here; you, the unconscious, or those you are involved with. Unresolved psychological issues, conflicts, or current personality patterns can be stumbling blocks or barriers to success. You must investigate and understand those factors which hold you back from achieving your maximum potential. Psychological insight into yourself, your own behavior, and the manipulations of others will be essential in having power over your own life.

Pluto indicates a confrontative style, and in the house of self it ensures confrontation with the deepest self. All power issues arise from self-awareness and the recognition of either the ability or inability to assert personal needs and rights when interacting with others. To reclaim power over yourself, you must start with psychological insight into your own unconscious blockages. You should not rest easy while standing in your own way, allowing your fears to inhibit your success. You must investigate if you are to overcome inhibitions blocking freedom of action and meaningful encounters with others. This is a time to bravely enter what you would normally consider dangerous territory. We all have areas of the mind or patterns of behavior that seem difficult to understand and frightening to reclaim or correct. These blockages are built and maintained by fear born of past situations which were handled ineptly. These difficult situations occurred at least once, but often a traumatic pattern with an inept response was repeated. Now is the time to learn to deal with these situations successfully and undo the blockages and trauma that have stood in your way too long. If you are able to gain insight into the fear motivating the inhibitions, you can now repeat a past crisis successfully and undo the complex. If you eliminate fears, inhibitions and the complexes they support, you can reclaim personal power and respond rationally to future events.

Your past trauma or current blockage is activated by a present-day situation; consequently, your psychological idiosyncrasies tend to be obvious to those in the immediate environment. They will undoubtedly know what sets you off, and may use this knowledge to gain power over you. Using an extreme example, suppose you have a compulsion to gamble and eventually incur huge gambling debts. Your inability to curb your own habit leads to an excessive dependency on money. Others can use this monetary need and gambling compulsion to their advantage. If they love you, they will use this information to help you, but if they are unscrupulous, they will use it to control you. You need to learn more about your psychological make-up in order to deal with self-defeating behavior and free yourself from bondage to others.

Ignoring the issues only makes you powerless. The struggle you experience with self and commonly associate with others is only an external manifestation of an internal process or blockage. You must look inside yourself for the answers. Events that occur during this year will affect you deeply, and may cause a permanent personality transition to new levels of awareness, self-expression and interaction. The potential is there for a milestone year. But the potential is also there for increased psychological pressure resulting from difficulties you refuse to face. Own your own power; don't give it away.

The task is to begin making conscious decisions rather than reacting from the gut level. If you refuse to face your own issues and claim what is rightfully yours, present circumstances and those you must deal with can have power over you. What you do not claim as your own can be claimed by lovers, friends and even enemies. You can choose to avoid the issues innate to your psyche, but then you also avoid the wonder of the power you possess and your ability to provide insight to others. The process of gaining control over yourself leads to power in all other areas of life.

This is the year to make a stand, and you are likely to confront others. Once you have healed yourself and owned your own power, you have the ability to give insight to others and influence their behavior also. The psychological power associated with Pluto can be used to heal more than one person. It can be used for your own healing or that of others. In the ideal situation, both healing processes occur simultaneously, with each individual involved contributing pieces of information leading to wholeness.

Be mindful not to control your environment unnecessarily. With Pluto in the 1st house, you may have a tendency to manipulate others in an attempt to divert attention from your own real issues. Even the facilitation of someone else's growth is suspect unless the facilitation is mutual. When in doubt, use insight to your own best advantage. Transform yourself first and then you will have the key needed to help others. You are the number one priority as illustrated by the placement of Pluto in the 1st house.

There are many avenues leading to transformation. Personal insight and vulnerability are but two. Manipulation and control are avoidance behaviors meant for diversion. The key to growth is to interpret everything that is said and done within the personal context first. Every experience drawn during the year will correspond to a personal need or lack. This is an excellent time for in-depth relationships and personal observation. Whatever the task for the year, it will be fueled by a need for intensity. Superficial or casual experiences will not suffice. Only power confrontations and intense relationships will have the depth of feeling necessary to foster transformative change.

PLUTO IN THE 2ND HOUSE

Dramatic changes in financial situations are associated with Pluto in the 2nd house. Your income may increase or decrease during the year and it is not uncommon to either enter or exit the job market at this time. Check the aspects from Pluto to 6th or 10th house planets for more information. In general, it appears that trines from Pluto to planets in these houses can indicate an increase, while squares tend to show the loss of income. Income losses can be preplanned and may not indicate any difficulty. You may wish to quit your job or retire. Unexpected salary cutbacks or financial difficulties are possible, but generally changes in salary come from self-initiated decisions.

The task for the year is to work towards controlling your own finances, including both income and outflow (spending practices). In the more positive manifestation, you will want to manage your own money. You must be the person in control, the one who decides how much you will or will not earn, and how you will spend it. It's time to either draw up a budget or trim the fat off the old one. Perhaps you want your own checking and/or savings account if you do not already have these. Learn to handle money responsibly. The tendency with this placement can be to maintain tight control over expenses. All purchases can be well thought out in advance, with allocations for specific expenses.

The need for personal financial control might cause a problem for your spouse, parent or significant other if oppositions run from the 2nd house to a planet in the 8th house. But it is also possible that this opposition simply denotes the changes occurring in shared finances as you move towards greater independence and control. Financial struggles over debts and expenses are possible, especially if you have not maintained good control in the past or need to tighten your budget now. You may need to make some changes in order to meet your goals for the future. Changes generally involve spending less and saving a sum of money for a large expense further down the road. Some individuals realize that they must be free of financial concerns now in order to freely pursue goals which are not financially rewarding at this time. For this, you must stockpile funds now and learn to live on less money.

Financial control may be an all-or-nothing deal. It may be the total lack of control that rules your life. In this case impulse spending and large expenses will drain your capital. The inability to control spending results in serious financial disruption. In this negative case, disruption will last for the year.

Changes in self-worth are frequently associated with changes in income, since many measure their self-worth against a materialistic yardstick. For them, money equals self-value. Equations such as this reinforce psychological messages which define personal worth in the context of conditional love. Those with poor self-images will not be able to support their sagging self-esteem without external monetary confirmation. Once income stops or the flow of money tightens, one must face the issue of innate value separate from abilities and accomplishments. Grow to appreciate who you are regardless of what you are earning or doing.

PLUTO IN THE 3RD HOUSE

The power of this placement lies with a psychological understanding of the workings of the unconscious mind as it relates to verbal communication. It is especially important to observe consistencies and inconsistencies between unconscious complexes and conscious thought patterns. Communication during the year is not intended to be one-dimensional, since what is meant is probably more important than what is said. Those who continue to focus their attention solely on conscious thought or word will lose insight into themselves, others, and what is really being communicated. New understanding gleaned from discussions is now meant to include psychological awareness. In order to grow and learn, one must begin to catch any discrepancies between what is actually said and body language or action taken by the person making the statement. Eventually one will become sensitized to most unconscious messages whether obvious or not, consistent or inconsistent.

This is not the time to deal with superficial discussions of the cocktail party variety. You will need to talk at length and in depth about important and "gutsy" topics. It is during these conversations that the unconscious complexes are most likely to manifest. Since it is sometimes difficult to obtain objective insight during conversations with family members and friends, this might be a good time to see a counselor or join a support group.

Power lies in the underlying message which is meant to program the listener to respond in a particular way. Words have power; however, it does not lie in the words themselves, but rather in the total message that is meant to be expressed and responded to. Sometimes the words are meant to reinforce a message and at other times they are meant to directly contradict the intended message. In any case, power and understanding are gained every time communication is truly understood. Perhaps an example might make this concept clearer. Suppose you wish to travel alone for the first time in your life and you plan a long vacation overseas. Part of the purpose of your trip is to overcome your fear of being by yourself. Your roommate of many years may respond in several ways. If he or she is truly happy for you and wishes to see you grow in self-confidence and control, the verbal messages you receive will be very supportive and the body language consistent. If he or she feels neglected or angry about not being invited to come along, you can receive very negative messages about your trip. Manipulative tactics or even threats may be used. If he or she is not in touch with unconscious anger and disappointment, but verbally appears to be positive, supportive messages sprinkled with warnings and fears about traveling alone may be the order of the day. In each of these possible scenarios, the underlying message conveyed by your roommate affects you psychologically by either augmenting or undoing unconscious complexes associated with independent travel and aloneness.

Fatedness can be a product of the unconscious mind. What you do not know or cannot face about yourself can control your behavior. Free-willed choices can result from the conscious, rational mind working with the unconscious mind to bring understanding and consistency to both facets of thought. Enlightenment leads to freedom of movement, and during this solar return year, power over destiny is closely associated with power of thought. Positive ways to increase your understanding of the unconscious include studying psychology (and body language), joining a discussion group, or regularly writing down your feelings and thoughts. Without this kind of focus, some individuals will still naturally move towards psychological awareness, but it may take an ongoing disagreement to do so. Practicing a positive technique may help you to avoid conflict altogether.

In manifestations involving conflict, manipulations associated with power struggles over intellectual concepts, behavior, or decisions might take place on a daily basis. Someone can be intolerant of your new ideas or obsessed with an antiquated line of reasoning. Negotiations and discussions will not be straightforward since unconscious complexes will complicate communications. Power plays and psychological ploys are common. Gossip may be a problem and you must be mindful of your reputation. Spiteful comments, whether truthful or fictitious, can be used to undermine your effectiveness. It may be necessary for you to stubbornly adhere to your own convictions in order to prevail or survive.

As the year progresses, it will be more and more obvious to you how the unconscious plays a role in everyday life situations. You will become aware of how you are being manipulated and how you, in turn, manage to manipulate others. The interpretation is not meant to appear one-sided and the manifestations of your own unconscious mind play a major role in the learning process. Personal complexes surface along with obsessive and/or compulsive tendencies. Although you are striving for a greater understanding of your own emotional and unconscious attitudes, at times you might feel more controlled by them than in control yourself. Major developments during the year might result from an unconscious need to undermine conscious decisions. Life may make a fated turn. It also becomes increasingly easier to dwell on one issue and allow it to rule your life, thoughts and moods. This is especially true if you are angry. Repressed anger can cause you to lash out at inappropriate times and for insignificant reasons, with reactions overtaking rational thoughts, fears dictating responses. A mind this receptive to stimuli may be so strongly influenced by another as to be subject to control by that person. This is why awareness of communication is so important if one is to retain power over self. In very negative situations, mental stability is questioned and therapy is indicated. Counseling intervention or consultation is common with this placement.

This is a great year for intense learning situations, even if the subject matter is not of a psychological or emotionally introspective nature. Any field of interest can stir a compulsive need to learn. During this period, you will not be satisfied with superficial explanations. You will strive to know and understand the underlying principles.

PLUTO IN THE 4TH HOUSE

Pluto in the solar return 4th house is the single best indicator of moving from one home to another. Uranus in the 4th house can also indicate relocation, but when Uranus is in the solar return 4th you are more likely to have changes or disruption within the domestic environment, particularly involving family members or roommates. Pluto, on the other hand, is more representative of moving from one home to another, or major renovations to the living structure itself, along with a disorientation or upheaval that lasts for a period of one year. These are fine-line distinctions, and of course, variations will occur. If you purchase a home during this year, it may need a lot of work. Redecoration is likely, and the repair of unforeseen problems is a possibility. It is in your best interest to have an engineer inspect any home you are seeking to purchase. Hidden or unanticipated difficulties can then be made known before any transaction takes place.

Relocation complications can arise for a number of reasons, though, and are not limited to repair problems. If you are building a new home, construction can take longer than expected. You may have to live in several temporary residences or stay with relatives for a period of time. Once you finally move in, you must decorate from scratch, purchasing everything an established home is already likely to have. Moves over great distances are very complicated in and of themselves, and generally require a full year's time for preparation and settling in. The hassles of adjusting to a new state, climate, culture, or environment can take the place of problems with the house structure itself.

Major renovations are also seen while Pluto is in the 4th house. These kinds of renovations necessarily cause a prolonged upheaval in the domestic situation. Large additions, modernizations, or constrution which involves moving walls or piping is likely, and the disruption caused by the dust and building will probably last for most of the solar return year.

If you do not move or renovate, it is still likely that changes in the domestic situation will occur. Commonly, someone either moves in or out, and living arrangements must be adjusted accordingly. New situations will take a while to get used to. All changes involve the consideration of many different factors, and simple decisions are not likely. In addition, transitions are likely to be complicated by psychological issues, and may even be unconsciously motivated. For example, your mother moves in and you must make arrangements for her to get to the senior citizen center during the day. She is somewhat incapacitated and needs daytime care. She also requires a living space on one floor and you must rearrange rooms and put in a new bathroom for her use. Psychologically, you must make the adjustment from living alone to caring for an elderly parent. Certain childhood issues might arise as you search for new patterns of behavior.

It is true that complications and struggles are associated with domestic situations when Pluto is in the 4th house; however, the complications and struggles are inherent in the project or transition you wish to complete and not the planetary placement itself. This is a time when individuals naturally seem to choose to make major changes in their home and/or living style, and it is unlikely that a transformation of this caliber could be made quickly and easily, without some hassle.

The desire to make strong and sweeping changes in the physical home is accompanied by corresponding emotional transformations. Sometimes the two manifestations are directly connected; sometimes they are totally separate; commonly, they are at least symbolically related. For example, during a long period of domestic upheaval, you could feel disconnected or not grounded, without a sense of "home," or a base for operations. If you are living with others, caught in a limbo between residences, you might be forced to subsist for a while without your own space, possessions, privacy, personal control, or sense of organization. You might feel stripped to the essence of your being, devoid of any external trappings, left to exist as you are, without the familiarity of what you own or can accomplish. Without the inner sanctum of your home for protection, you could feel exposed, and it is true that you are vulnerable to the idiosyncrasies of temporary housemates. Emotions are bound to fluctuate under such conditions. Old behavior patterns break down quickly as new patterns and coping mechanisms are necessitated by changing circumstances. In this way, Pluto in the 4th house usually indicates an emotionally unsettled time, as well as a physically unsettled time.

Sometimes this emotional upheaval can occur because of a coincidental physical change in another's situation, but a transition in the native's own consciousness is essential for growth. Suppose your parents decide to sell your childhood home and move out of state. If you are still attached to your parents' residence, you must make adjustments in your thinking. You can still visit your parents, but you can never go home again if the trip involves one particular building or location. Perhaps it is time for you to think of your own residence as "home."

When Pluto is in the 4th house, emotions are not only the motivating force behind physical changes, but also the end result. The purpose and desire associated with this placement is the need to establish an emotionally fulfilling environment, both on the internal and external planes. This is why moving or changes in the physical home (external) are as important as evolutionary growth in the feeling nature or unconscious (internal). The two levels of manifestation go hand in hand during the year, and you can be sure that if your physical surroundings are in a state of transition, corresponding changes are also affecting your emotional level.

By the same token, the inability to make wanted and needed external changes probably indicates the presence of internal blockages and external resistances. A search is underway to form stronger roots in the physical and emotional environment. Shallow connections will not do. One must either dig deeper or move on; reaffirm the commitment to the existing structure (renovation), or begin anew (move). The same process is occurring on the emotional level. Individuals must reaffirm a commitment to family members and roommates, or seek support elsewhere.

Unlike other 4th house placements which also indicate changes in the physical and emotional home, Pluto generally indicates that not only are changes long overdue, but there are also some inner and outer resistances which must be overcome to fully realize the internal potential in the external world. Consequently, a power struggle with oneself and others is likely, both before and during change, and the arena for this struggle takes place in the home and among family members. Conflicts arise when others do not want the same home environment you envision. In the best of circumstances, negotiation and compromise may be all that are necessary to resolve differences. However, in some families the word "negotiation" is really a cover for manipulative tactics and controlling influences. Nothing is ever resolved, issues remain unsettled, and decisions are not finalized. Within this domestic milieu, the creative urge towards rootedness and its resultant manifestations on the physical and emotional planes are stifled before they can be realized. For this reason, you should consciously work to undo old, existing response patterns which allow others to psychologically control your emotional nature and rob you of its creative potential. Being emotionally controlled is generally synonymous with also being physically controlled, and therefore unable to create the home and home life you desire.

This is a good time to enter therapy, especially if unconscious manifestations (fear, guilt, negativity, power struggles, etc.) are blocking new and old commitments, or stifling your urge towards growth and change. Repetitive issues, unfulfilling emotional interchanges, and the inability to cope with your home and necessary changes signal the need to reassess response patterns for power leakages. During the year, heightened sensitivity will cause you to become aware of psychological games which rob you of personal power in the emotional arena. At this time, even an intellectual interest in psychology can give you great insight into domestic and familial patterns.

Because of the issue of creative emotional power and its manifestation on the physical plane, life seems more crucial, decisions more important, problems more complicated. Everything leads to something else and you have a vested interest in how matters finally turn out. Commonly, the meaning of life is discussed, and within this climate the issue of death arises, causing you to face the issue of your own mortality. There may be a death in the family during the year, but more often you are merely acquainted with someone your own age who must deal with a potentially life-threatening illness. His or her situation calls you to consider the possibility of your own death, and the reason for your existence. The interplay between the creative emotional nature and the issue of mortality leads you to understand the extent and limitations of your own power. One does not have control over death; all living things die. However, one does have power over the emotional response to death, and for that matter, all emotional responses. The need to direct your own emotions towa a rewarding and fulfilling expression is at the root of this Plu placement.

PLUTO IN THE 5TH HOUSE

Pluto in the 5th house commonly indicates a power struggle ov self-identity and self-expression. Unconscious messages around yo can be working to mold your persona into a reflection of somebod else's personal needs. As the year begins you suddenly realize th you are not solely the product of your own creation; someone else exerting pressure on you to perform in a particular manner. The pus for performance may or may not be in your best interest, but th ultimate goal for you to accomplish during the year is creative sel mastery. You must learn how to balance the external demands with the internal push for self-expression. Freedom to be your own perso is crucial to growth. Two external forces are at work here and th messages to the unconscious are twofold. First there is the externa definition of what one is meant to be. Messages commonly play o existing unconscious forces, such as fears, and therefore are accept easily. After this message is inculcated, it is followed closely by a pus for conformity. Because of this dual mechanism one can be partiall controlled by others, since self-expression will be limited by one own automatic reactions to unconscious messages, and by susceptibility to psychological forces exerted by another person.

For example, one man constantly received messages from h boss about how he needed to respond in order to be successful. Th boss implied that not being a "company man" and a "team playe meant never working in the business again. The boss's messag played on the man's own fears concerning inadequacy, and h readily began to play the role of the dedicated employee. But then th job demands became excessive. The hallmark of the negative mani festation for this placement is the eventual emergence of excessiv demands on the part of the person seeking control. For our youn man, somewhere between the unconscious messages and the pu for conformity, he began to lose his own self-identity. Numerou hours of overtime robbed him of any personal life.

It makes no difference who sends the messages to the unco scious. It may be a lover, a child, or another; realistically, it can b anyone you respond to at an unconscious level. Sometimes the lines of stress shown by the aspects to Pluto can denote the individu triggering the issues related to power over self-expression. Often it is an intense relationship which brings out facets of your personality normally hidden and subject to control. Negative relationships can be manipulative, but very positive relationships can encourage growth in this area. Responses can be healthy and not all situations are bad. For example, a mother responding to the demands of her newborn child is conforming to his or her needs. An intense and somewhat controlling relationship develops between the two, which in this context, is considered normal. Becoming aware of the controlling process and working towards balancing external demands with the internal need for self-identity can lead to self-mastery. By understanding the psychological pressures, you can choose to conform to those which are positive experiences while refusing to conform to those which seriously jeopardize self-identity. In very difficult situations, the inability to balance the conflict between expectations and self-identity can result in more secretive behavior. You may prefer to do things on the sly rather than weather the confrontation necessary for true self-expression.

Generally, intense relationships are needed to spark the push for self-definition and personal control over self-expression. You will meet many people over the coming year who can either threaten or augment your ability to accomplish this task. Love relationships, especially new attractions of a Plutonian nature, have the intensity necessary for this process. The overpowering compulsions associated with these relationships usually force the individual to redefine the boundaries of self-identity by looking at the unconscious responses which prevent the self from being fully empowered.

Although sexual attractions and love affairs are not always seen with this placement, when they do occur the draw is tremendous and the allure of sexuality seems especially compelling. The person you are attracted to and your relationship with him or her will bring out the best and the worst you have to offer. Insights into your own motivations can be overwhelming and you will be acutely aware of when you are and when you are not in control. Sometimes you will not be totally rational. The relationship itself creates intense feelings which can defy logic and overpower common sense while giving great insight. Unconscious messages, fears and expectations might surface during simple everyday interactions, making it clear that you are not able to entirely control your own life, your involvement with this person, or your commitment to him or her. If you are free to pursue a love affair, you will learn a lot about yourself and your ability to give and receive sexual pleasure. If you are not free, the relationship will still be both compelling and conflicted while accentuating the need to address unconscious complexes.

Pluto in the 5th house can indicate a birth during the year. The intensity of the relationship between a newborn and its parent is consistent with the need to maintain some sense of self-identity while meeting the demands of the child. As a new parent with this placement, you may not be used to understanding all the effects children can have on those who raise them. Babies are little individuals with personalities of their own which may or may not fit your expectations and needs. Right from the moment they are born, they have the ability to elicit psychological responses from others and all parents must conform somewhat to the needs of their children. Infants demand the greatest amount of attention, but even as your children mature they will retain an ability to affect you and your psychological state. When they grow old enough to talk, they might question any conscious or unconscious inconsistencies you manifest. As teenagers, they can buck your authority. As adults, they may choose lifestyles which you find psychologically stressful. Whatever their age, there may be little you can actually do to control their behavior until you analyze the psychological responses coming from your own unconscious. If your children are young, you must gain some measure of control, but realize that complete control at any age is unattainable. You must bend and accommodate to a certain degree, while maintaining a sense of self-identity. Know which hang-ups are your own. Learn to make your own responses more appropriate to the issues at hand. You can easily overreact if the situations make you feel overpowered.

New patterns of creativity can evolve, especially if you are in a field where psychological and emotional issues can be incorporated into your endeavors. Issues associated with conflict, power and intense love might manifest in creative work. Artistic creations tend to exhibit more depth of expression, and major stylistic changes are possible. Mundane rather than artistic projects can also be highly creative. Usually tasks and projects foster personal power and increased expression of the inner self.

PLUTO IN THE 6TH HOUSE

Pluto in the 6th house of the solar return implies changes in both the daily work situation and personal health practices. Changes in physical health are also possible, but generally not as likely. In regard to employment, your work environment could change dramatically. The most common manifestation involves an office which either totally reorganizes (e.g., computerization), or moves to another location. This causes a great upheaval in the daily routine until everyone learns the new procedures, or figures out where everything is located. Preparation, implementation and resolution time may take almost a year. In lieu of this, you alone may change bosses, or departments, or be given a new assignment or job description. Daily working conditions are likely to become more complicated until the period of transition ends.

This is a good time to eliminate unnecessary daily tasks. Obsessive-compulsive tendencies may have you doing more work than is necessary. You can get hung up on details, or feel pressured to stay on the job longer than your normally scheduled hours. In these situations, work begins to have power over you, and you are no longer in control. You can become a workaholic this year if you are not mindful of the need to balance your workload. If you analyze your productivity, you may discover that your time is not efficiently organized. Learn to streamline your day by creating more efficient daily routines. It is possible that tasks can be completed in less time and with less effort. Take corrective action in those problem areas. If you are self-employed, hire someone to help you with the clerical work so you can be free to work on other projects. If you are working for a company, develop a plan of action and present it to your employer or manager. This is a time when even the lowliest employee will seek some power within the work environment. Positive use of this desire can lead to improvements on the job.

Power in the workplace becomes an issue, and you can use personal power as a lever for success. For example, one employee with a wealth of valuable information negotiated a higher salary and a position of authority. He was then able to use his position and abilities to transform a failing business into a successful enterprise. On the more negative side, power struggles with co-workers or lower level managers are possible. Someone may have a lot of power over you, watching what you do, when, and how. You may feel like you must be on your guard. Backbiting, gossip, and underhanded manipulative tactics can be the norm if you choose to participate. Undercurrents and back-room maneuvers dictate policy. Sources of contention are difficult to discern or confront since nothing is ever truly out in the open. Power over your own schedule or working habits might be totally out of your control as you are forced to conform to inane rules. If you wish to quit, you are likely to do so. Differences of opinion, conflicts with authorities and disputes over the implementation of new ideas can make it impossible for you to continue in the same position. If, however, you choose to remain on a job you sincerely hate, be forewarned that health problems will most likely arise.

You are capable of having a lot of control over both your physical and psychological health; however, early in the year it may not seem so. Instead, you may be struck by how compulsive your habits have become. A few of you may be locked into serious addictions to alcohol, drugs, or cigarettes, but most will only experience the need to control a craving for a particular junk food, or a resistance to an exercise routine. During the year, poor health habits can be a problem and usually these habits do not arise suddenly; rather, they are long-standing patterns which only now demand corrective attention. The push for control is indicated by Pluto's presence in the 6th house. This is a good time to make conscious, rather than unconscious, choices about health practices. You can break addictions or bad habits, especially if you also treat the underlying psychological issues.

It is important to understand that your state of health or disease is directly related to your psychological diet. The more you are in tune with your inner self, the better you will feel; the more you are in touch with your environment and the people around you, the more likely you are to feel emotionally nourished. A healthy emotional climate breeds a healthy mind and body. On the other hand, the greater the stress in the environment and the more you are suppressed or manipulated by others, the greater the chance of health problems arising. Emotional upsets can directly affect your health, especially if you are caught in "damned if you do and damned if you don't," no-win situations. The emotional diet is as important as the nutritional diet, and even though you take care of yourself physically, mental or emotional stress or abuse can make you ill. If you are in a difficult situation, consider these three options. You can walk away from unending conflicts or spiritually unhealthy environments. Secondly, recognize psychological games and refuse to be manipulated. Protect your unconscious from damage. Thirdly, use relaxation techniques and spiritual insight to alleviate stress. Realize your power to grow healthy, both inside and out.

PLUTO IN THE 7TH HOUSE

While Pluto is in the 7th house, issues concerning relationships become complex interactions which must be analyzed to be fully understood. Awareness is being raised to a new level of understanding and there is no book or course you can take which will give you the information you need for this passage. Knowledge springs from the inner reaches of the mind, and is fueled by the compulsions and frustrations realized in both intimate and nonintimate relationships.

We must make a distinction here between two different relationship processes, and three different levels at which you may choose to work. The two different processes are love and hate, and the three different levels of interaction include nonintimate partnerships, intimate relationships, and soul-level attractions. (All three will be defined and explained.) It makes a difference whether you choose to learn through lessons of love or hate. Certainly the information gained will be different; however, the compulsion to see remains the same. Those who choose to see through love excel despite despair, while those who choose to hate will despair despite their ability to excel. The issue of power is innate to this placement and cannot be taken lightly. One must acquire and maintain personal and relationship power through new insight and understanding. Power may be acquired through love as easily as through hate; the choice is yours, but generally those who seek to love gain power over self, while those who hate (control) seek to gain power over others.

Nonintimate relationships consist of everyday acquaintances and business partnerships. At this level, almost anyone can trigger the need to face the issue of power as it flows through relationships. Conflicts are either specific or diffuse; one person in particular may become the trigger to powerlessness, or a power deficit may exist in all relationships in general. For example, you may be locked into a business partnership in which each individual is trying to meet certain goals and needs. This type of struggle relates to a specific area of your life, and the intensity is mostly restricted to that area, though insight and tension may affect interactions with other people.

Generalized power struggles are caused by an inability to actualize personal power in any relationship. In this case we are not dealing with one particular situation, but with a diffuse problem which encompasses all interactions. The inability to be assertive is the most common manifestation of this nature. Both types of struggles, whether specific or generalized, cause external confrontations which force the individual to look at the unconscious need to maintain power while interacting. This need can be met through an increased understanding of the mechanisms which govern the power flow.

Psychological issues are crucial with this Pluto placement. It is at the nonintimate level that one first begins to understand how psychological motivations and fears affect the way one communicates and relates. Blatant manipulation can be the primary form of expression, in which case much of the struggle will be nonverbal in nature. You may be the instigator or you may be the victim; it makes no difference since each position correlates with psychological complexes you must understand and conquer. There is never the one-sided attack; all struggles represent a mutual process whereby the aggressor either knowingly or unconsciously triggers reactions in another. In the most negative manifestation, power struggles are difficult hate battles that last most of the year. Legal confrontations are possible. Enemies can arise and some attacks seem unwarranted. But those who master the psychological influences create new patterns of relating which represent power delicately balanced. Old partnerships, regenerated, become cooperative. Even those with generalized assertiveness problems can learn to express personal needs to others.

Intimate personal or family relationships lead the individual to make further distinctions in the understanding of psychological influences as they affect relating. External conflicts may be very apparent (similar to the nonintimate level), but it is only in intimate relationships that daily exchanges can produce the subtle insights necessary for understanding psychological complexes at a deeper level. It is within this context that old psychological problems such as obsessions, compulsions, addictions, jealousies and control issues tend to surface full-blown and begin to play a much larger role. Personality traits and idiosyncrasies also affect the ability to relate in a meaningful manner. During the year, your relationship with another, usually a lover or spouse, will go through a period of transition. Both of you must look at the mechanisms by which you relate. The desire to control another is usually a central theme with manipulation and game-playing inherent in the process. You need to be aware of these ploys since they are impediments to greater intimacy and freedom. Complex power dynamics involving love and hate eliminate the possibility of freedom for one or both partners. Each must [illegible] as the other bids in order to suppress the unconscious fears motivating the need to control.

It is only through insight and an understanding of the underlying fears that one is able to dismantle the psychological complexes and begin to handle relationships clearly. The year can bring greater intimacy to those who are willing to work together to strengthen their commitment to one another while at the same time dismantling the control mechanism. Because of the placement in your solar return, you must make the transition, but it is up to others to decide whether or not to also make the necessary changes. If you are in need of a deeper, more intimate relationship than your loved one is capable of at this time, you may seek counseling, sever your present commitment, or seek other avenues for intimate exchanges. There are no easy solutions to the complex problems of relating.

If you are not already in a relationship, this can be a milestone year for you, one in which you are strongly attracted to someone new or someone you have been previously only acquainted with. Lost loves may return. Intimacy needs are increased at this time and you now need in-depth encounters. Intensity will be the norm and you do not care to waste time on superficial interactions. Even nonromantic relationships can have an overpowering effect on you. Your psyche is vulnerable to the insight of others. Even those you meet only briefly can have a tremendous effect on your life.

Soul-level manifestations involve new relationships which are karmic attractions that force the individual to seriously question all past and present relationships. The triggering mechanism is a desire for a new level of intimacy. The person you are drawn to may not be representative of someone you would choose for yourself on the conscious level. The implication here is that the unconscious chooses and there is no room for cliched romances. Only something very different will create the intensity necessary for the overwhelming growth pattern associated with this placement.

The issues that are dealt with this year involve a serious challenge to your ability to handle intimacy in a new way. The questions one should ask when faced with a relationship of this intensity are, "To what depths am I willing to go in order to acquire the insight necessary to understand my relationships as they exist now? Am I willing to face myself truthfully as one-half of and contributor to a complex interaction that affects both my capacity for soul growth and also the ability of others to excel? Am I willing to pay the price of vulnerability and honesty to acquire the highest level of intimacy to which I may aspire?" Intimacy at this level and intensity demands that one dismantle all defenses, and stop all ploys for power. Tr power comes from shredding the persona to reveal the true self. doing so, one gains power over self and encourages all others to let go of useless power ploys also. It is at this point that meaningful relating on a karmic level can begin.

PLUTO IN THE 8TH HOUSE

It is very common for financial arrangements to change during the year. If you live alone and are your sole financial support, the amount of money you earn could either increase or decrease. T same is true if you receive money from parents or other relatives. If you share resources with another person, either one of you can decide to quit your job or cut back on hours. Rarely, one may lose his or job. A raise is also possible, but budget adjustments are most likely in either case. Debts are an issue and you may endeavor to wipe out all debts. In instances where one person is the major spender, he or should be asked to personally cover the bills. The debt situation may have grown out of control, in which case it could be time to cut up your credit cards. If you are in serious trouble, bankruptcy, fore sure and poor credit ratings might ensue.

In actuality, though, the amount of money is generally not the issue; control is more important. If you live alone and are your o financial support, you will not want to be controlled by money or the lack thereof. This is not the time to stay in a job strictly for financial reasons. This is not the time to spend wildly without any sense of control. Financial power struggles are associated with this placement, and if you live alone and are your sole support, the struggle for control could be with yourself. In other situations, the power struggle may occur with your employer. You could feel that your employer is not treating you fairly; salary inequities are possible, or you may be asked to sign an exclusivity contract, limiting your ability to moonlight and earn outside income. If you live with another person (spouse, lover, or roommate), monetary struggles between the two of you are possible. Who is or who is not controlling the money, and how it is allocated, become important issues. The two of you may have different ideas about how money should be earned or spent. Sometimes demands are made or money is withheld. In very negative situations, manipulative games are played in which money is traded for sexual favors and financial blackmail occurs.

Sexual harassment is a rare yet possible occurrence, since the struggle for power can occur in almost any situation. You might be harassed by a professor, employer or fellow employee. It might be very important for you to defend or define your sexuality during the year. Overbearing personalities can confront you or block your progress. Sexual prejudice and sex-role stereotyping rob you of your effectiveness in competitive circumstances and are ultimately just another struggle for power and control.

Difficulties with insurance companies and insurance claims seem to be associated with this placement. Individuals have reported canceled policies, unpaid claims, disputes over coverage, and general mix-ups. If you have dealings with insurance companies during this year, allow extra time to work out conflicts or errors. Don't be afraid to fight for what is rightfully yours, because you can win.

These struggles all emphasize the importance of financial control rather than the amount of money. Struggles can sometimes occur over very small amounts when the principle of the thing is involved. The underlying concept is the need to reassess and control your own monetary situation, especially when other people are involved. Money can be handled more equitably. You must have power over your own financial future, and dividing certain expenses in a fair way may be the best answer. In very difficult circumstances, you may have to assume all financial responsibility. To gain any control, one must be aware of psychological influences affecting the financial situation. Spending practices and monetary power plays are more likely to be motivated by unconscious complexes than by logical reasoning. If you are psychologically insightful, your understanding of the issues associated with this placement will be more complete.

Psychological information is important generally, and not just in regard to financial and sexual situations. The study of astrology, psychology or any other esoteric or occult practice can also be important. Information and knowledge of a less obvious nature is indicated by this placement. Perception is increased, though others may not be aware of your new insight. You do not have to rely solely on what you are told. For example, you could become adept at reading body language. You grow to know more than you are told and see more than you are meant to see. In positive situations, your newfound skills are used for healing. You are able to correct or modify difficult behavior patterns in yourself and others during the year. In very negative situations, you are faced with intense psychological confrontations involving guilt, resentment and manipulative tactics. The issue of control surfaces in the psychological arena and you have to deal with unconsciously motivated actions, someone else's or your own. Association with aberrant personalities might occur at some point during the year, but more likely behaviors will remain within normal limits, though therapeutic intervention might be required. The focus of attention may be internal rather than external if a push for self-knowledge becomes the focus of your new insight. If others are using your own unconscious complexes to control you, you can dismantle the manipulative games through counseling. You need to limit your own unconscious contribution to power struggles with others. Self-knowledge and learning to stay at a rational level of communication will enable you to regain power over yourself, your involvements and your situation.

PLUTO IN THE 9TH HOUSE

An intense learning period can occur while Pluto is in the 9th house, and this is regardless of whether or not you are actually in school. Generally it is not the academic environment which is most notable, but the obsessive interest in one particular topic, and the philosophical shift which is caused by the influx of much new information. Like a good book you cannot put down, your newfound interest may cause you to skip work or neglect chores. Learning becomes both a time-consuming process and an overwhelming influence. Your belief system will be strongly affected by what you learn.

During the year, you can take a home-correspondence course, study on your own, or attend a formal school. Problems at school are possible and they can range from minor inconveniences to major difficulties. Those in a structured learning environment may have to cope with classroom disruption, or curriculum complications. The school building itself could be in the process of renovation. School policy changes could affect you or your course of study. Classes you want or need can fill up early, forcing you to take courses in a nonsequential order. Building renovations, policy changes and scheduling difficulties are usually minor inconveniences. More difficult problems involve the disruption of your ability to concentrate and do your best. You may be psychologically affected by circumstances at school which make it hard for you to study or respond rationally. For example, you may have very positive or negative feelings toward a particular professor which inhibit your ability to contribute to classroom discussions. The atmosphere in the dormitory is probably not conducive to study, and if you are used to make quiet a period of adjustment will be necessary. Your relationship with your roommate, boyfriend or girlfriend may break your concentration. If you disagree with school policy, you could choose to buck the system or become a disciplinary problem. All of these facts and more can influence your educational experience. For those who have been in a traditional school or college up until this time, this probably is the year you leave the academic environment for one reason or another.

If you are a teacher or lecturer, some of the above might apply, but you are more likely to see disruption in the 10th or 6th house because it is your employment. If you disagree with school administration policies, or if you are in the process of rewriting those policies yourself, both 9th and 10th house placements are likely. This is a time when you can be very forceful concerning your beliefs, and your attitudes probably carry over into the professional arena.

Radical philosophical changes are likely, especially if your beliefs are directly challenged by someone during the coming year. Within the academic environment, you will be challenged by what you are learning. Within a social setting, interactions with people of different religions, cultures, or ethnic backgrounds will challenge your preconceived ideas and prejudices. Familiarity with these foreign customs or religious practices may be especially helpful in furthering your understanding of different lifestyles. But a philosophical confrontation is not essential for you to change. Even everyday life situations can directly contradict a long-held belief or perspective on life.

One factor consistent with this placement is that the effect on the belief system tends to be stronger and more noticeable if you are emotionally involved with the very person or people who cause this philosophical revision. The impact of an emotional or psychological connection makes the conversion, attack or insight much more effective. The new information you are learning can drastically change your beliefs and make you feel disoriented until you have adjusted your philosophy accordingly. Do not lose sight of self. Remember to maintain a personal perspective. Beliefs about yourself are just as likely to be affected as beliefs about others. Opinions you retain may have to be defended against the attacks of others. Challenges are meant to force you to review and solidify your position.

At issue here is a philosophy of life. Belief systems have a controlling influence on our behavior, and many times they form a basis for operation. To be at your best, you must be clear about what you believe, and beliefs must be an accurate representation of your reality. A constant interaction between your experience of the world and your understanding of that experience leads to a healthy philosophy. Feedback systems and constant review are important. If what you believe to be true is directly contradicted by your experience, and you still continue with the same mind-set, this could be a very difficult time for you. Tightly held misconceptions can eventually cause psychological abnormalities. Reality is a strict teacher.

Some individuals are more controlled by their philosophies than served by them. These people are subject to prejudicial attitudes and intolerance. Their belief systems have no continuing basis in reality and therefore they must close their minds to any new information or interaction which is contradictory. These people respond in an overbearing or belligerent manner because they cannot afford to listen. Fanatical tendencies are possible. The whole purpose of this Pluto placement is philosophical change brought about through new insight. Belief systems were meant to be dynamic in the first place, but dramatic shifts are likely during this year. If you can step back, listen, and observe without preconceptions, much will be learned. Insight and wisdom can be gained by those who accept the challenge to either defend or revise beliefs, but the greatest advances are made by those who stop to listen first.

PLUTO IN THE 10TH HOUSE

Pluto in the 10th house is indicative of a strong career or destiny transition. Important decisions are made during the year, and these decisions will have a lasting effect. The tendency is to come to a fork in the road, and your psychological response to the options offered sets the pattern for future growth.

On a mundane level, most of the emphasis centers on a career push which may or may not involve a change in profession or employment. Some individuals change or lose jobs, while others forge ahead in their present positions, acquiring power and authority through promotion. Generally, changes are occurring on external and internal levels, so both job and attitudinal changes are likely. If you own and operate your own company, it is time to reassess your goals for the future, and the way you do business in the present. Certain practices or structures can limit your future business needs. You might choose to expand, streamline or shut down.

Pluto rules births, and deaths, and complicated processes. Problems must be addressed, and whatever you plan to initiate, the transition will probably evolve over a year's time. If you are employed by others, you might wish to change jobs, especially if old professional pursuits are no longer fulfilling. On the other hand, you can instill new life into your old job. Introducing new concepts, tasks and philosophies to your manager or co-workers can lead to a changed job description, or future career possibilities. Ambition is implied by this placement, and advancement is possible for those who are willing to work for it, or are able to negotiate the corporate power structure. A compulsive push for success now will enable you to make career moves which propel you to a higher professional bracket. Use ambition to hang in there for the long haul and finish those long-term projects which are crucial to advancement.

Career difficulties are also associated with Pluto in the 10th house. Sometimes the push for success ends in a power struggle for advancement or a manipulative conflict for psychological supremacy. Controlling power is an issue here, and sometimes the emphasis is on a negative exploitation of position and authority. You can be the victim or the perpetrator; most likely you play both roles. An obsession with advancement breeds workaholic tendencies and an unhealthy working environment. Within this climate, power plays begin to take precedence over promotion as manipulative tactics disrupt normal business. Just as hard work and healthy ambition push you towards success, psychological complexes and power struggles can thwart your efforts. Within this arena of struggle, conflicts with an authority figure are possible, and not just those in the business world. Government officials, probation officers and sometimes parents fall into the authoritarian category. If you are in a middle management position, how you handle authority can be as important as how well you follow directions.

Power issues are not limited to the work environment but will arise in all areas of life. The type of power you are attuned to will be representative of the area of awareness you are working on. Some will be made aware of physical force, others psychological control, still others will focus on spiritual power. Each level of insight can correspond to an awakening. Although the 10th house is the house of career, it is also the house of destiny, and of those issues and choices which will have a strong effect on the destiny path.

It is during this year that some are given a destiny choice. They may continue in their current pattern of growth and behavior, or they may make a leap in consciousness. There is a fork in the road and a decision to be made. You will not pass this way again. The opportunity is now and the only thing standing in the way is fear. Options are so far reaching that they produce remarkable changes which might force old situations and relationships to fall away or completely transform. The magnitude of the decision will not escape you. But the choice is there for those who wish to go on to greater things. The destiny change is made through the use and awareness of power on one level or another. A choice of this caliber can be made in the career arena where one learns to effectively use power in the business world, or the choice can be made in the personal arena. For some the choice concerns the use of spiritual power. Choices and the effective use of power affect the pattern of growth for many years to come.

PLUTO IN THE 11TH HOUSE

Pluto in the 11th house commonly implies involvement in a group and the experience of power and transformation within the setting. You may be the leader or a follower, but in either case you must handle the issue of power as it relates to individual needs versus group needs. This can be a time when you accomplish more through group efforts than you do on your own. On the other hand, you might feel that your goals are compromised by a shared effort. Individual differences between various members and the distinction between the needs of the one and the needs of the whole become more apparent. One can't always do it alone, just as one can't always be the corporate player. For this reason, group power struggles are likely, and not only do you witness the struggle, usually you are a part of the interplay within the hierarchy. Sometimes you are able to make your way, learning to handle power gracefully. Sometimes you feel the need to fight for what you believe in. Other times you choose to withdraw entirely. Situations can become very complicated and there are no easy answers to issues that arise.

Psychological influences affect group interaction, and for this reason they can have an intense transforming quality. If you can catch the moments of manipulation within the power plays, you can learn about yourself, others and group dynamics. Motivations are never totally altruistic, and one should look for the real reason certain events are occurring. Look within as well as without. Circumstances which make you feel out of control are especially important, for within them is a power leakage. Truly difficult situations can leave you feeling drained and emotionally overwhelmed. A more positive possibility for this placement could be joining a self-help or encounter group which engages in intense in-depth conversations meant to stimulate insight.

All of the above interpretations can apply to friends or one friend in particular, as well as to groups. A special friend who is willing to risk intimacy can have as strong an effect on you as an encounter group. Power struggles are also just as likely when expectations are involved. One woman depended on her friends to help her in a new business and this became an issue when they could not always show up. She sought to press them into service by equating support with loyalty.

Changes in goals commonly occur. As you develop new ones, old ones are given up. What was once feasible is no longer realistic; what is new shows great potential. Highly specific and very ambitious goals are possible, and this is the route many individuals take. The possibility of never completing tasks is very real. Difficulties in attainment are associated with psychological blockages and power struggles. When this is the situation, life seems too difficult or complicated to allow for the formation or attainment of goals.

The urge to direct power at a specific goal can be shortsighted for those who are capable of more. The true wonder of this placement does not lie in the attainment of any goal, but rather in the process of individualization. A specific goal is finite; the recognition of your own ability to be different, an individual on your own terms, is infinite. Many use the specific attainable goal to initiate the process of awareness, but others tend to stop there without doing the real work that needs to be done. As long as you use power only for an attainable goal, you sell yourself short. You lose even more power if you allow yourself to get locked into a power struggle. You have the right to be your own person, so the freedom needed to manifest your individuality need not come from strife. It is already yours. Power struggles represent a need to establish an individual identity. Some can do this on their own; some only through the experience of otherness. Some work with; some work against. You choose the path of awareness, but the basic need is to recognize and develop the power that lies within your own individual differences.

PLUTO IN THE 12TH HOUSE

The 12th house rules things that are hidden, behind-the-scenes and/or not easily recognized. When Pluto, the planet associated with unconscious and hidden forces, is in the 12th house, the interpretation is particularly subtle, but also especially important. What is happening behind-the-scenes may be more important to growth in consciousness than what is happening in the public environment. Use this time for backroom negotiations if it is to your advantage not to be very public about what you are doing. For example, you might be designing and testing a new product, or having a love affair. The power of this placement lies in what you know, but do not say or make others aware of. A son comforted his father during his illness with massage and holistic techniques. Because the father had strong religious beliefs, the son never talked about what he was doing or the holistic concepts behind the techniques. It was enough to comfort his father. This can be a time to deal with behind-the-scenes issues in a quiet, rather than confrontative, manner.

A more negative interpretation might emphasize manipulation and underhanded tactics rather than a positive use of discretion. Motivations are not necessarily noble. Undercurrents might be ignored and not discussed. Sometimes you are not aware of the problem yourself. If Pluto doesn't cross the Ascendant by transit into the 1st house during the year, a hidden focus will remain; however, if Pluto crosses over, the personal urge will be to reveal or deal with what was previously hidden.

The same is true with major changes that are waiting in the wings. When Pluto is in the 12th, but close to the Ascendant, there is usually a change of significant magnitude being formulated. It is more likely to occur during the year if Pluto actually crosses. If Pluto comes close but never actually transits into the 1st house, look in the following solar return chart for an indication of change. You may not quite make the change, or only partially make the change this year.

Psychological complexes are very important with this placement, but they are also less controllable and more difficult to recognize. You may try to ignore your own issues. It is easy to point the finger at someone else and declare that the other person has the problem without being aware of how you, yourself, contribute to their difficulties with your own psychological idiosyncrasies. Compulsive tendencies are common. For example, you may feel compulsively drawn to a new love relationship. There is an interplay here between the compulsion, your urge to control the unconscious, and the frank realization that this is hard to do. The past comes back to haunt you, and present situations may be of a *deja vu* nature, making control especially difficult. If Pluto stays in the 12th, you are less likely to confront your own psychological complexes; however, if it crosses, you should feel the need to do so.

The higher concepts associated with Pluto, or any planet for that matter, are not always readily understandable to the individual. There are levels and degrees for all interpretations. You must work for the higher manifestation and there is a price to pay. You cannot be focused on personal power or gain and be in touch with Universal consciousness. Personal motives and true spirituality are the antithesis of one another, like oil and water. The oil (personal gain) covers the water (Universal consciousness) so that it cannot reach the surface and is drowned out by the oil. If you try to reach down for the water, your finger will come up greasy and barely wet. Personal motives must be put aside to truly understand the Plutonian process and work with it in its purest form. The point of focus for Pluto in the 12th is the ability to draw Universal power through the unconscious mind. The free-association and visualization capabilities of the unconscious are the perfect vehicle to enlightenment and growth. Those who are pure of heart and mind and soul become universal receptors, receiving information that needs to be on the earth plane. For this, there is no need for compensation or glory; it is enough to know.

CHAPTER TWELVE

GENERAL INTERPRETATION AND DOUBLE CHART TECHNIQUE

The interpretation for the solar return chart can stand alone and does not have to be related to the natal chart. A wealth of information can be obtained from just the solar return chart by itself, the house placements of the planets and the aspects between them. But for those who look for natal confirmation or who want to compare the solar return to the natal chart, you can place the natal chart outside the wheel of the solar return. In many cases the natal planets denote the driving forces behind the solar return interpretation. Think of the solar return as a play put on by the natal planets. Every year there is a new script, complete with new scenery and new roles.

The script is loosely written, and really only introduced by the solar return interpretation. It leaves plenty of room for improvisation in the outcome and process. The roles or personality traits are signified by the solar return planets, while the scenery is defined by their house placements. With a 4th house emphasis, scenes take place in the home. With planets in the 10th house, the play centers on the career. A variety of planetary placements in the solar return is associated with complex scene changes. The relationships among various issues, traits and conflicts are illustrated by the aspects within the chart.

The natal planets are innate traits, and therefore they are the actors in the play. They take on the roles shown by the solar return placements. All the actors have personalities of their own which either complement or detract from the roles they are trying to assume. It is possible that the natal planet (actor) will be asked to play a role for which it is not well-suited or prepared. For example, if your solar return Venus is conjunct Uranus in your 5th house of love affairs, but also conjunct your natal Saturn, the interpretation becomes very complex. The Saturnian concepts of structure and caution might inhibit a wild and carefree love affair; however a tendency towards balance might be advantageous.

Natal planets symbolically radiate through the solar return planets and give some coloring to the parts they must play; however, the most obvious issues during the year are associated with the solar return planets themselves and their placement. Where the natal characteristics shine through is shown by the aspects from the natal planets to those of the solar return. One natal planet might aspect and color the interpretation for several solar return planets. It is important to assess the themes associated with a natal planet and how those motifs might affect the psychological traits and issues reflected in the aspected planets of the solar return.

Other considerations in interpreting the solar return and natal chart together include the orientation of the house cusps. The natal house which covers the solar return Ascendant is important in the analysis of the background themes which are the motivating forces behind the year's activities.

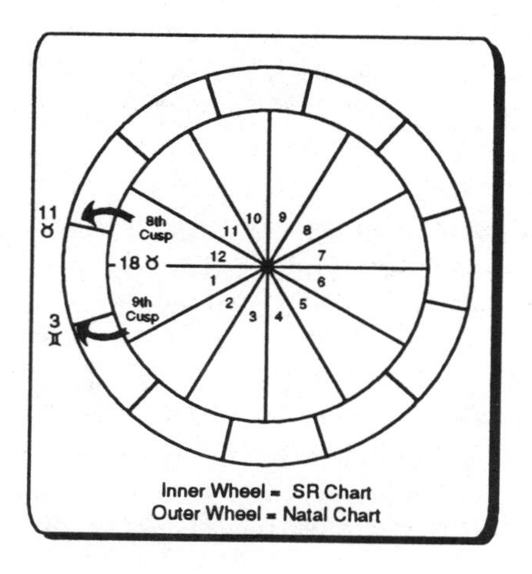

Figure 25

For example, if the solar return Ascendant is 18 Taurus and t natal 8th house cusp 11 Taurus (natal 9th house cusp 3 Gemini), t solar return Ascendant will fall into and be covered by the natal 8th house. This could mean that shared resources or money from spouse or partner could be a motivating force during the coming ye It may be that the spouse provides money needed for a project or venture which is shown in more detail within the solar return ch itself.

The natal house covering the solar return Ascendant is made more obvious by a natal planet also conjunct the Ascendant. Contin ing with our example, suppose the natal 8th house Jupiter is at Taurus.

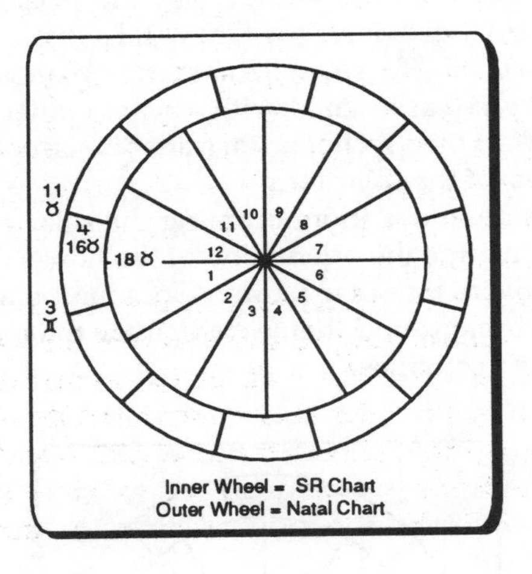

Figure 26

This would again support the possibility of receiving money perhaps in a lump sum from others. The interpretation of the natal planet conjunct the Ascendant directly relates back to that plane natal house position.

The house covering the Midheaven can give you an idea about the motivations behind career events and choices. As with na conjunctions to the solar return Ascendant, natal planets conjunct the Midheaven should also be prominent in your interpretations, with the meaning relating back to the natal planet house position.

Interceptions between the natal houses and the solar return houses show some form of limitation, and usually the limitation is directly related to the intercepted house. For example, a person with the 7th house of the natal chart intercepted in the 2nd house of the solar return might feel that it is untimely to get married during the year since funds are inadequate. Interceptions are commonly perceived limitations and not necessarily actual external restrictions. For example, if your 8th solar return house is intercepted in your 2nd natal house, you might decide to limit spending and not incur debt. As a rule, it does not matter if the solar return house is intercepted in the natal house or vice versa. The tendency will still be to set boundaries or limits on activities.

For those who really want to get into the fine points of interpretation, the combination of any natal house covering any solar return house can be assigned a meaning. For example, the 2nd house covering the cusp of the 5th house might indicate that your financial situation is affected by your children's needs. Traditional ethics might influence the way you raise your children or conduct a love affair. Insight can be gleaned from any of the cusp combinations between the two charts.

By interpreting the relationship of the solar return to the natal chart, one can confirm insights and also gain further information if needed. But remember, the bulk of the information concerning yearly issues lies within the solar return chart itself rather than the aspects (relationships) between the solar return and the natal chart. The issues are defined in the solar return. How you will tend to respond—to handle those issues—is implied by the natal chart.)

CHAPTER THIRTEEN

TIMING TECHNIQUES IN THE SOLAR RETURN CHART

There are several timing techniques that can be used with the solar return chart. The two most useful seem to be the progress solar return Moon in aspect to solar return or natal planets, a transiting outer planets over the solar return house cusps.

The solar return Moon can be progressed in a manner similar that of the secondary progressed Moon. Notice the speed of the Mo on the day of the Sun's return and divide this travel distance by twelve to arrive at the monthly movement. The Moon should moving approximately one degree per month in the solar return sin it moves approximately eleven to fourteen degrees per day in the ephemeris. Once you have calculated the monthly speed of the Moo, you can add this figure to the solar return Moon position to come with the progressed monthly placements which will aspect both solar return and natal planets at specific times during the year. Aspe made to solar return planets and angles are generally more noticea than aspects made by the progressed solar return Moon to natal planets and angles. The only exception might be if the progress solar return Moon conjuncted a major natal configuration. Then t interpretation would tend to be as obvious as an aspect to the solar return planets themselves, rather than secondary.

Transits of major planets to the solar return angles and ho cusps generally work nicely. They can even be used to rectify the natal birthtime. We do not have major transits crossing our natal Ascendants, Descendants, Midheavens and 4th house cusps very frequently, but because of the rotation of the solar return chart from year to year, we are more likely to experience this conjunction. The timing of a transit over an angle can be very helpful.

For example, a woman who was employed and planned to quit her job did so only after eliminating extra bills. She felt an obligation to pay off all family debts before she stopped bringing in a second income. Saturn was in the 12th house of the solar return chart and she wanted to assume responsibility for her actions. On her last night of employment, Saturn conjoined her solar return Ascendant and crossed into the 1st house. Another woman bought her very first lottery ticket and won $250.00 during the time transiting Jupiter crossed the Ascendant of her solar return chart.

Planets crossing the Ascendant are especially important, if they are outer planets poised in the 12th house. Usually there is a coinciding event at the time of the transit. If the outer planet never actually crosses the Ascendant, events are more likely to stay in the planning stage for the moment or become near misses. Transits to the other angles, Descendant, Midheaven and 4th house cusp are equally significant.

Transits to the solar return house cusps are also important, though not as noticeable as transits to the angles. The movement from one house to another lends a secondary interpretation to a planet's meaning in the solar return chart. You can interpret the planet in both houses once it has crossed over into the later house. For example, if Jupiter is poised in the 1st house, but conjunct the 2nd house cusp, you can create a self-made opportunity which will bring you a lump sum of money as Jupiter crosses into the 2nd. Because solar return planets just inside a house and conjunct a cusp have just transited in, they tend to indicate events that occurred just before the Sun's return. On the other hand, if a planet is poised just outside of a house, but conjunct the cusp within a degree or two, it can indicate that a event or psychological shift will occur shortly after the Sun's return.

Other timing techniques which can be helpful, but tend not to be so consistent or noticeable, are the Sun transiting the solar return planets, and retrograde planets conjuncting their solar return positions. The transiting Sun conjoing solar return planets can be the symbolic trigger for events, especially if other more significant transits are occurring within the same span of time. It is also important to note when a retrograde or direct solar return planet travels back and forth over itself in the yearly chart.

Solar returns are meant to be used in conjunction with secondary progressions, solar arcs and/or transits to the natal chart when one is trying to time trends. The techniques mentioned above can be very helpful in analyzing information from several techniques and combining it into a cohesive package; but, taken by themselves, they cannot be as accurate as a combined perspective.

CHAPTER FOURTEEN

CLOSING THOUGHTS

Solar return charts are guides for those who can work with a yearly study plan. In actuality, though, because of your experience, you are still last year's solar return and all the previous ones. Anyone who has ever had a certain house placement in their solar return can easily draw on the power and consciousness that has come before. The former stage of growth is always with you since life experiences are never lost, only taken to a new level of understanding. When a planet's house placement returns for the second, third, or fourth time, this signals a need for further growth in those areas. Situations at this time foster new insights and a deeper understanding of life, but the new awareness is always an extension of what was previously understood or misinterpreted.

Solar returns represent an accumulative process, wherein one year is built on another. The more you accomplish this year, the more you will be capable of accomplishing next year. As skills are attained, talents flourish and options multiply. The goal during any one year should be to work with your present situation, while still drawing on past knowledge and expertise, and preparing for future challenges. Retention and preparation are as important as being in the here and now. Use that which you were, to be who you are, and become what you will be.

The process can be taken one step further. Insights fostered by your present circumstances can be integrated and transferred across the chart and across various life situations. Pluto in the 7th of relationships can manifest in all the houses for those who are enlightened enough to choose to relate issues from one house to another. For example, insights picked up through relationships can influence the way you communicate to others (3rd house), or the career decisions you make (10th house). Translations can follow the aspect lines in the solar return chart, but need not do so. We sometimes limit our awareness because of astrology. We see a planet, in a sign, in a house, making aspects, and fail to comprehend the limitlessness of growth and consciousness. In reality, none of the planetary placements is limited to a house position for those who see beyond. There are no real limitations, and should not be. The house position of a planet is merely the starting place, a springboard to new understanding, the initial arena of experience. Every chart should be seen as a beginning, not an end. Your experience of the symbolism of any planet can expand across the chart, becoming a comprehension of the whole and all the integral parts, planets, placements, and aspects. Use the solar return to focus your attention on the initial areas of consciousness seeking awareness, but then use the information and insight obtained to foster growth multilaterally across the chart and across all life situations. Consciousness is limitless for those who are willing.

Index

Boldface type indicates major discussions.

N

O

P

Q

R

S

T

U

V

W

X-Y-Z

No entries

Also by ACS Publications

All About Astrology Series of booklets
The American Atlas, Expanded Fifth Edition (Shanks)
The American Book of Tables (Michelsen)
The American Ephemeris Series 1901-2000
The American Ephemeris for the 20th Century [Noon or Midnight], Rev. 5th Ed.
The American Ephemeris for the 21st Century 2001-2050, Rev. 2nd Ed.
The American Heliocentric Ephemeris 1901-2000
The American Midpoint Ephemeris 1991-1995
The American Sidereal Ephemeris 1976-2000
Asteroid Goddesses (George & Bloch)
Astro-Alchemy: Making the Most of Your Transits (Negus)
Astro Essentials: Planets in Signs, Houses & Aspects (Pottenger)
Astrological Games People Play (Ashman)
Astrological Insights into Personality (Lundsted)
Basic Astrology: A Guide for Teachers & Students (Negus)
Basic Astrology: A Workbook for Students (Negus)
The Book of Jupiter (Waram)
The Book of Neptune (Waram)
The Changing Sky: A Practical Guide to the New Predictive Astrology (Forrest)
Complete Horoscope Interpretation (Pottenger)
Cosmic Combinations: A Book of Astrological Exercises (Negus)
Dial Detective: Investigation with the 90° Dial (Simms)
Easy Tarot Guide (Masino)
Expanding Astrology's Universe (Dobyns)
Hands That Heal (Burns)
Healing with the Horoscope: A Guide To Counseling (Pottenger)
Houses of the Horoscope (Herbst)
The Inner Sky: The Dynamic New Astrology for Everyone (Forrest)
The International Atlas, Revised Third Edition (Shanks)
The Koch Book of Tables (Michelsen)
Midpoints: Unleashing the Power of the Planets (Munkasey)
New Insights into Astrology (Press)
The Night Speaks: A Meditation on the Astrological Worldview (Forrest)
The Only Way to... Learn Astrology, Vols. I-VI (March & McEvers)
  Volume I - Basic Principles
  Volume II - Math & Interpretation Techniques
  Volume III - Horoscope Analysis
  Volume IV- Learn About Tomorrow: Current Patterns
  Volume V - Learn About Relationships: Synastry Techniques
  Volume VI - Learn About Horary and Electional Astrology
Planetary Heredity (M. Gauquelin)
Planetary Planting (Riotte)
Planets in Work: A Complete Guide to Vocational Astrology (Binder)
Psychology of the Planets (F. Gauquelin)
Roadmap to Your Future (Ashman)
Skymates: The Astrology of Love, Sex and Intimacy (S. & J. Forrest)
Spirit Guides: We Are Not Alone (Belhayes)
Tables of Planetary Phenømena (Michelsen)
Twelve Wings of the Eagle (Simms)
The Way of the Spirit: The Wisdom of the Ancient Nanina (Whiskers)
Your Magical Child (Simms)
Your Starway to Love (Pottenger)